Introductory Principles of Plant Breeding

Second Edition

W0080819

Introductory Principles of Plant Breeding

Second Edition

RC Chaudhary MSc (Ag) Bot, PhD

Former
Associate Professor, GB Pant University of Agriculture and
Technology Pantnagar, Uttarakhand
Professor of Plant Breeding and Regional Director
Rajendra Agricultural University, Pusa, Bihar
Rice Specialist, The World Bank, Nigeria
Plant Breeder and INGER Global Coordinator, IRRI, Philippines
Chief Technical Advisor, and Project Manager, FAO, Rome Italy

Currently Chairman, PRDF, Gorakhpur, UP

Oxford & IBH Publishing Co. Pvt. Ltd.
New Delhi
(*A Unit of* CBS Publishers & Distributors Pvt Ltd)

CBS

CBS Publishers & Distributors Pvt Ltd

New Delhi • Bengaluru • Chennai • Kochi • Kolkata • Mumbai
Bhopal • Bhubaneswar • Hyderabad • Jharkhand • Nagpur
• Patna • Pune • Uttarakhand • Dhaka (Bangladesh)

Introductory
Principles of
Plant Breeding
Second Edition

ISBN-13 978-81-204-1775-5
ISBN-10 81-204-1775-5

OXFORD & IBH
New Delhi
(A Unit of CBS Publishers & Distributors Pvt Ltd)

Published by Satish Kumar Jain and produced by Varun Jain for
CBS Publishers & Distributors Pvt Ltd
4819/XI Prahlad Street, 24 Ansari Road, Daryaganj, New Delhi 110 002, India.
Ph: 23289259, 23266861, 23266867   Fax: 011-23243014   Website: www.cbspd.com
e-mail: delhi@cbspd.com; cbspubs@airtelmail.in.

Corporate Office: 204 FIE, Industrial Area, Patparganj, Delhi 110 092
Ph: 011-4934 4934   Fax: 011-4934 4935   e-mail:publishing@cbspd.com;
                                                        publicity@cbspd.com

Branches

- **Bengaluru:** Seema House 2975, 17th Cross, K.R. Road, Banasankari 2nd Stage, Bengaluru 560 070, Karnataka
  Ph: +91-80-26771678/79   Fax: +91-80-26771680   e-mail: bangalore@cbspd.com
- **Chennai:** 7, Subbaraya Street, Shenoy Nagar, Chennai 600 030, Tamil Nadu
  Ph: +91-44-26260666, 26208620   Fax: +91-44-42032115   e-mail: chennai@cbspd.com
- **Kochi:** 42/1325, 1326, Power House Road, Opp KSEB Power House, Ernakulam 682 018, Kochi, Kerala
  Ph: +91-484-4059061-65   Fax: +91-484-4059065   e-mail: kochi@cbspd.com
- **Kolkata:** No. 6/B, Ground Floor, Rameswar Shaw Road, Kolkata-700014 (West Bengal), India
  Ph: +91-33-2289-1126, 2289-1127, 2289-1128   e-mail: kolkata@cbspd.com
- **Mumbai:** 83-C, Dr E Moses Road, Worli, Mumbai-400018, Maharashtra
  Ph: +91-22-24902340/41   Fax: +91-22-24902342   e-mail: mumbai@cbspd.com

Representatives

• Bhopal	0-8319310552	• Bhubaneswar	0-9911037372	• Hyderabad	0-9885175004
• Jharkhand	0-9811541605	• Nagpur	0-9421945513	• Patna	0-9334159340
• Pune	0-9623451994	• Uttarakhand	0-9716462459		
• Dhaka (Bangladesh)	01912-003485				

Printed at Rashtriya Printers, Dilshad Garden, Delhi, India

Preface to the Second Edition

The book with the title 'Introduction to Plant Breeding' was first written in 1980 primarily to serve as a textbook for Under Graduate students of Indian universities. Due to its simplicity and to the point and well explained text, it became so popular that dozen times it was reprinted. Those days, loaded with the concurrent responsibilities of the jobs at The World Bank as Rice Specialist (1984 - 1989), and then with International Rice Research Institute (IRRI) as Plant Breeder and as INGER Global Coordinator (1988 - 1997), I could not find time to revise the book. Immediately thereafter with Food and Agriculture Organization of the United Nations (FAO) for one and half decade followed and pressure on my time remained the same. In 1980s most of the available books on the subject were of American and European origin, and covered irrelevant topics and examples. Student had to go through several books to complete the course requirement. Later several books written by Indian authors appeared. Still my book remained popular with the Professors and students alike. Even during the year 2013 i.e. 33 years after its first publication in 1980, thousands of copies were sold. That was an eye opener and encouragement for me to revise and expand. Moreover, after retirement from FAO, I am relatively free and could devote some quality time on it. While consulting various professors about the course outline, I realized that following last Dean's

Committee Report, course name and outline was also was changed from Introductory Plant Breeding to Principles of Plant Breeding. Various agricultural universities were allowed freedom to make variation in the course content to the tune of 5%. All these variations had to be included to help students in getting required knowledge for this course. Then plant breeding is dynamic subject and activities of plant breeders keep adding many new methods and varieties of different crops. New agricultural universities also appeared on the horizon. All these events warranted serious revision and updating of the book, which has been done in the current revision. Thus the information has been updated but keeping the basic structure and simplicity of the presentation at the same level. A new chapter on Innovative Methods in Plant Breeding has been added.

Special help received from Dr. Stephen Nielen if IAEA for a consolidated list of mutant varieties; Dr. C. Aruna and Dr. J. V. Patil of Directorate of Sorghum Research Hyderabad; Prof. H.S. Chawla, Prof. J.P. Jaiswal of GBPUAT Pantnagar, Prof. N.K. Singh and Prof. V.K. Chaudhary of RAU Pusa Samastipur, Dr. Ashok Kumar Singh of IARI New Delhi, Prof. J.L. Durvedi, Prof. Prem Kumar and Prof. R.K. Srivastawa of NDUAT Faizabad Prof. Dinesh Yadav and Dr. Vanplana Kashyap of Gorakhpur and many others is gratefully acknowledged here.

The publisher M/S Oxford & IBH gladly accepted the proposal and cooperated that within the shortest time this revised edition is in your hands. Appreciate any one more if I receive your comments and suggestions for further improvement.

While surveying the course coverage, help rendered by Heads of Departments of Genetic and Plant Breeding / Agricultural Botany of various universities is gratefully acknowledged. Information's and photographs supplied by all is acknowledged gratefully. My wife, Mrs. Rukmini Chaudhary, has given constant inspiration and fullest cooperation in completing the revision. She deserves all appreciation for the help and moral support that she rendered.

Ram C. Chaudhary
Gorakhpur   Ram.Chaudhary@gmail.com
February 2014

Foreword

Plant Breeding has been described as an art as well as a science. A successful plant breeder is generally well trained in both the art and science of plant breeding. This text on Introduction to Plant Breeding is written to introduce the undergraduate students of Indian universities to both the facets of plant breeding. It covers the subject matter is most logical and sequential manner. Elementary principles of genetics, which are the foundation stones of plant breeding, are explained in precise terms. There are excellent discussions of breeding methods and the factors, which determine the choice of breeding methods for a particular crop. The role of special breeding methods such as mutation breeding, heterosis breeding and polyploidy breeding is adequately reviewed. The chapter on breeding for disease and insect resistance deals with this important subject in a lucid manner. There is a brief discussion of achievements of plant breeding in some important crops and release of varieties and seed production in the last two chapters.

The book abounds in examples of plant breeding work done in India. The varietal names of Indian crops used in the book would be familiar to the students of Indian universities. Because of the pivotal role played by the improved crop varieties in increasing the food production in the country, more and more students are being attracted to major in plant breeding. The publication of this book is therefore timely and it should prove useful both for the teachers as well as students.

I would like to congratulate Dr. R. C. Chaudhary for the commendable job of preparing this text. I am confident it will prove extremely useful and will receive wide acceptance.

Plant Breeder and Head,   **GURDEV S. KHUSH**
Plant Breeding Department
International rice research Institute
Los Banos, Laguna, Philippines

Preface to the First Edition

The book 'Introduction to Plant Breeding' has been written primarily to serve as a textbook for undergraduate students of Indian universities. Most of the available books on the subject are of American and European origin. They cover a wide range of subjects, yet a student is required to go through several books to complete his course requirement. Besides, examples cited in those books are not easily grasped due to unfamiliarity with variety names and prevalent problems in crop. Some Indian books are available on the subject, but are not up-to-date. It was therefore thought that students should be provided a comprehensive book as per their course requirements.

While preparing this book it has been assumed that as a prerequisite, students have acquired fair knowledge of crop botany and elementary genetics. The chapters have been so arranged as to give a sequential knowledge of the science of plant breeding, methods of breeding in various groups of plants, specialised methods of breeding and finally the procedure of release and seed production of varieties so developed. A chapter on the history, institutional organisation and achievements in breeding various crops has been included. Up-to-date examples to illustrate various points have been chosen from different universities and institutions of the country. Thus, this book would serve not only undergraduate but postgraduate students of plant Breeding, Agricultural Botany, and Botany, of various universities, also as a source of revision.

While surveying the course coverage, help rendered by Heads of Departments of Plant Breeding/Agricultural Botany of various universities is gratefully acknowledged. Informations and photographs supplied by Dr. M. S. Swaminathan, ex- Director General, ICAR; Dr. N. G. P. Rao, Project Director Sorghum; Dr. R. N. Singh, Joint Director (Research), IARI; Dr. P.S. Bhatnagar, Project Co-ordinator Sugarbeet; Dr. H. K. Shama Rao of BARC, Trombay; Dr. I. D. Singh of TRA, Tocklai, Dr. T.R. Hargrove of IRRI, Philippines is gratefully acknowledged. Work of distinguished scientists and authors, which has been quoted, is duly acknowledged. Several of my colleagues at Pantnagar Dr. R.L. Agrawal, Dr. P. L. Gautam and Dr. U.K. Rai, Dr. Sunil Saran of Patna, Dr. B. Rai and many others of RAU have helped me in various ways, Dr. V. Sivasubramaniam, of TNAU, Coimbatore, Dr. H.K. Mohanty of OUAT, Bhubaneshwar, Dr. S.D.I.E. Gunawardena of CARI, Sri Lanka, and Dr. S. S. Virmani of IRRI, Philippines, gave many useful suggestions in the manuscript.

I am particular thankful to Dr. Dhyan Pal Singh, Vice Chancellor of Rajendra Agricultural University, Pusa (Samastipur) Bihar, for permitting me to prepare the book. My wife, Mrs. Rukmini Chaudhary, has given constant inspiration and fullest cooperation. She deserves all appreciation for the help that she rendered. Last but not least, I would be thankful to readers for pointing out errors and omissions, which might have crept in.

Patna   R.C. CHAUDHARY
1980

Contents

List of Figures

Plant Breeding
and its Scope

The sun is the ultimate source of energy for all living creatures on earth. Solar energy trapped by chlorophyll of the green plants, is the sole source of human food, fibre and fuel, either directly or indirectly. Thus, man has been concerned with plants since the beginning of the civilisation. With the development of science, the study of economic plants and their improvement has developed into a well-organised branch called 'plant breeding'. Ever since the beginning of agriculture, consciously or unconsciously, man has been creating genotypes, which are more efficient in trapping solar energy for converting it into food, fibre and fuel. Plant breeding is thus one of the earliest and foremost endeavours of man, which has led him towards civilisation and settled living.

1. DEFINITION

Plant breeding is concerned with developing varieties superior to existing ones. Frankel (1968) defined "*plant breeding is the genetic adjustment of plants to the social, cultural, economic and technological aspects of the environment*".

Several accomplishments of plant breeding can be cited to illustrate the above definition. The nutritional quality of several food plants has been upgraded by genetic means. For example, maize protein is poor in the essential amino acid called lysine. Probably this deficiency was responsible for the destruction of maize-based civilisations like Incas and Mayas of Latin America. Now, new maize varieties like *Protina*, *Shakti* and *HQPM* 7 were evolved by incorporating opaque-2 gene, which doubled the amount of lysine. Protein malnutrition is a curse to world population, particularly in developing countries. The protein content of major cereals, wheat and maize is being up-graded continuously. Breeding cotton varieties free of gossypol enables the use of cotton seed oil and protein for human consumption. This is genetic adjustment of crops to meet the needs of the society. Development of varieties with low neurotoxin in *Lathyrus*, low *trypsin* inhibitor in pulses, and low or zero *erucic* acid in rapes are the other goals and achievements of plant breeders. This adjustment upgraded the value of these crops to prevent nutritional disorders in human being. Golden rice rich in iron and vitamin A, Orange-Fleshed Sweetpotato varieties like *Sree Kanaka* and ST-14 with very high level of Beta carotene (Vitamin A) are produced by plant breeders to address health problems through biofortification.

Development of semi-dwarf varieties of rice and wheat, etc, has been primarily to enable the plant to respond to high doses of fertilizers and elevated levels of management, and yield more. Increased production named Green Revolution, mainly due to high yielding varieties, saved human population from starvation. Early maturing and photo insensitive varieties have been developed which fit into multiple cropping and various crop rotations. Development of fruit trees for high density orchards is yet another set of examples of genetic adjustment of plants for new cultural and economic arena. The mango variety Amrapali permits planting of twice the number of plants per hectare than older varieties. New apple varieties can be cultivated like tomatoes by staking and produce more fruit than wood. Thus plant breeders are able to produce tailor made designer varieties of plants to address any human need.

Under the prevailing socio-economic conditions in most developing countries, the agricultural technology most prevalent is the one, which involves labour intensive techniques. A good example of plant breeding under such conditions is the development of hybrid varieties of cotton, where hybrid seeds are produced by hand emasculation and pollination. The production of hybrid seeds in a hectare of cotton field provides employment for about 100 women for 80 days. Hybrid seed production of vegetables also set the same example of employment generation.

Now, here are a few examples to illustrate the genetic adjustment of plants to technological needs: development of a 'combine tomato' variety in the USA in which all fruits ripen at the same time to enable picking by a tomato combine. Similarly, in Australia pigeon pea varieties were developed which can be harvested with a combine harvester. Certain chemical treatments make some types of cotton cloth 'easy care'. Thus cotton varieties Hybrid 4, *Deviraj*, *Bhagya* and *Jayadhar* have been developed which are suitable for easy care treatment. Cotton fibre is white and to prepare coloured dress it is to be coloured with synthetic dye. Now cotton varieties with different colour like red, blue etc have been developed which does not need artificial colouring.

2. PLANT BREEDING AS AN ART AND SCIENCE

Ever since prehistoric man started cultivating plants, he first selected, consciously or unconsciously, superior plants from natural population. In the second phase he saved seed of his best plants for sowing next year. In a modest way this was the beginning of plant breeding. But there was no science in it; simply it was an art of selection. Plant breeding continued to be art up till the end of the 19th century when Mendelian Laws were rediscovered.

The modern technique of crop improvement developed on sound scientific basis only after the rediscovery of the Mendelian laws of inheritance. The contributions of other branches of science such as cytology, taxonomy, physiology, anatomy, biochemistry, biometry and now biotechnology towards the development of modern plant breeding technique can never be lost sight of. With the amalgamation of these different branches of basic sciences today, plant breeding has become a special branch of science dealing with evolving improved crop varieties and has been playing a pivotal role in the development of agriculture all over the world.

As the knowledge of plant genetics and related sciences progressed, the laws of inheritance were discovered and it became possible to reshuffle genes to tailor a desired plant type. Now with the biotechnological tools, it is possible to reshuffle genes among living organism. Thus, the art of plant breeding grew into a well knit science. The choice of parents based on knowledge of gene action, use of the most appropriate breeding methodology, analysis and interpretation of data are all science. Biometrical principles made even the art of selection fairly predictable. Therefore, as a science, plant breeding is only a little over 100 years old. But now plant breeding is more a science and less an art and involves more brain than brawn.

3. PLANT TYPE CONCEPT

Agriculture is the largest solar energy harvesting enterprise in the world. A part of the solar energy falling on the earth is trapped by green plants and converted into carbon compounds on which life of human race is dependent. Development of the plant canopy that can trap the maximum sunlight and produce maximum photosynthate vis-à-vis higher yield becomes an important consideration. This led to the concept of plant type or plant ideotype.

An ideotype can be defined as the required plant structure and developmental sequence of a plant which can best suit a particular environment for producing maximum economic return (Donald, 1968). Thus, the designing, breeding, testing and exploitation of plant ideotypes are logical steps towards new levels of yield. An ideotype provide a guideline to a breeder in the selection of the parents to be crossed and in the selection pressure to be imposed in the later generations. The ideotype may differ depending on species and systems of farming.

4. GOAL OF PLANT BREEDING

Providing a variety better than the existing one, has always been the aim of plant breeders. Thus, higher yield is and would remain the foremost goal followed by other attributes. Better quality in respect of taste, nutrition, suitability, industrial use, keeping quality, transportation and preservation-particularly in fruits, vegetables, flowers and food grains-are also important. Special quality in the produce, such as higher sugar recovery in sugarcane and sugarbeet; longer, finer and stronger fibre in cotton; more protein in pulses; more oil in oilseeds are other goals. Resistance to diseases, insect pests, cold and heat, flood and drought, adverse soil conditions and change in duration and suitability for new practices and possibilities are other goals. Environment resilient plant breeding is shaping the future course of this science.

Besides, any specific requirements of the consumer, the processor, the retailer and of course the nation, becomes the goal of the plant breeder. Naturally these goals are dynamic and keep on changing in time and space.

5. CONTRIBUTION OF PLANT BREEDING

5.1 Self-sufficiency in Food:

An English clergyman, Thomas Malthus, had predicted that unless checked by war or disease, the human population would grow so much that people would die of hunger by the end of the 19th century. This did not happen, and the main credit goes to agriculturists, particularly plant breeders, who met the food demand of the additional population by increasing crop productivity.

Similarly in the recent past India was classed among the countries which cannot be saved from starvation by any amount of world aid. Within a few years, however, the Green Revolution took place. Food production doubled due to the high yielding varieties of wheat, rice and other crops.

The production of food grains in 1977 grew so much that imports were stopped, old debts were paid and the country became self-sufficient. The Green Revolution triggered by high yielding varieties brought a complete change in production technology, marketing, storage and extension.

The productivity (kilogram per hectare) of all major food and fibre crops rose gradually 5 to 6 times (Table 1.1). This was all due to higher yielding varieties developed during the last 60 years in India. On the main strength of high yielding varieties, backed by area expansion and higher input use, production of food grains rose dramatically, from 61 million tons of 1950s to 230 million tons in 2012. India became net exporter of rice, wheat, maize and several other crops. Undoubtedly factors such as large area, better inputs, etc, played a role but high yielding varieties have contributed significantly to this quantum jump in production.

Table 1.1 Productivity (kg/ha) of major crops over last 60 years (Reserve Bank of India, 2013)

Year	Rice	Wheat	All cereals	Pulses	All food crops
1950–51	668	663	542	441	522
1960–61	1,013	851	753	539	710
1970–71	1,123	1,307	949	524	872
1980–81	1,336	1,630	1,142	473	1,023
1990–91	1,740	2,281	1,571	578	1,380
2000–01	1,901	2,708	1,844	544	1,626
2010–11	2,240	2,938	2,247	689	1,921
2012–13	2,462	3,119	2,450	786	2,125

5.2 Stabilising Production:

This can be achieved only by developing varieties, which can mitigate, to the extent possible, the adverse impact of aberrant weather, damage by pests, diseases and weeds. New varieties resistant to important diseases and pests are available in major crops, which indeed is a significant contribution of plant breeding. Resistant varieties do not require any fungicides or insecticides or other chemicals to protect them and thus environmental pollution is also avoided. Stabilised wheat production, for instance, is dependent on rust resistant varieties. In fact, the development of resistant varieties has been one of the biggest achievements of plant breeders to stabilize production.

5.3 Multiple Cropping

Development of short duration varieties of wheat, rice, pulses and oil seeds have made it possible to take 3-4 crops per year instead of 1-2 as in the past. Short duration urd (T-9) and mung (Pusa Baisakhi) varieties can now be cultivated in summer after wheat. Short duration pigeon pea varieties like Prabhat, Pant A-3, Bahar are harvested in six months instead of 11, and two crops can be grown instead of one in the same field. Many more new varieties of various crops are being developed which would fit in multiple cropping.

5.4 Tailoring Plant Characters

Plant breeders strive to achieve the best of all possible combinations of characters: plants with high yielding capacity, resistant to pests, weeds and diseases, tolerant to climatic stresses like cold, drought, water logging and submergence, amenability to local agronomic practices, resistant to storage pests and acceptable to consumers. Since most characters are governed by independent sets of genes, it is possible to recover, in crosses, segregants with desirable character combination. 'The plants of the world have a lot of wonderful genes if you can just find them and combine them', says Norman Borlaug. With the advent of the 'plant type' concept in plant breeding, plant structure has been so tailored to produce genotypes more responsive to better management. For example, traditional varieties of rice yield around 20 quintals but under better management, say 120 kg N/ha, would lodge and yield very low. The new dwarf varieties yield 60-80 quintals at 120 kg N/ha and under better management. The plant type of pigeon pea has been so changed as to make it smaller and more productive. Apple varieties have been developed at East Malling in U.K., which are not like old tall (100 m) and wood producing types but dwarf (1-2 m tall) and fruit producing types.

5.5 Extending the Boundaries of Cultivation

Newer varieties developed by plant breeders are adapted to wider areas and newer agroclimates. Cauliflower whose origin and domestication occurred under the temperate European climate has now yielded varieties, due to continued natural selection, which are adapted to tropical and sub-tropical conditions. These developments have taken place over a period of 40 years in India. Similar progress has been achieved in potato and tomato also.

5.6 More Nutrition per unit Area

Plant breeding has accomplished one very significant landmark; developing varieties which would produce more nutrition per unit of area. For example, the early sugarbeet varieties had about 4 percent sugar while there are varieties with 19 per cent sugar now. Multiplied by the higher root yield, the production of sugar per unit area has risen by 10 times. Similar is the example of oil production by some oil seed crops like safflower and protein by pulses. Since arable land area cannot be increased with the growing population, increased productivity becomes imperative and this is an important contribution of plant breeding.

6. TRAINING OF A PLANT BREEDER

Since a plant breeder is basically a biologist, he requires training in morphology, anatomy, cytology, systematic botany, embryology and physiology, particularly of crop plants. The knowledge of genetics is basic to him/her as the principles under lying plant breeding are basically dependent on the laws of inheritance. S/he should also acquire knowledge about evolution in crop plants and their domestication. A competent plant breeder has also to be a competent geneticist.

A plant variety must have resistance against pests and diseases. To develop a resistant variety, the breeder must have thorough understanding of plant pathology and entomology.

Statistics has helped plant breeders to compare large numbers of advanced lines, construct selection indices, and predict their performances. This area has developed into the biometrical approach to plant breeding. This is why in professional plant breeding, biometry is being emphasised.

Since experimental lines, breeding populations and materials for evaluation have to be grown using standard techniques. Hence knowledge of crop growing practices, and harvesting and processing technology is essential.

A plant breeder has to be trained in the nutritional quality of plant products since s/he is basically involved in upgrading the nutritional quality of the produce by genetic means.

In short, a plant breeder has to be a versatile person with a broad spectrum of knowledge of genetics, biotechnology, plant protection, crop husbandry, human nutrition, statistics and farm machinery. S/he is not necessarily expert in all these but should be in a position to apply the knowledge and experiences of the above disciplines in developing varieties.

7. USEFUL REFERENCES

Allard, R.W. 1960. *Principles of Plant Breeding*. John Wiley & Sons Inc., New York

Anon 1972. Genetic Vulnerability of Major Crops. NAS, Washington.

Anon. 2013. A Handbook of Agriculture. 6th edition. Indian Council of Agriculture Research, New Delhi

Balint, A . 1970. Protein Growth by Plant Breeding. Akademiai Kiado, Budapest.

Briggs, F.N. and Knowles, P.F. 1967. *Introduction to Plant Breeding*. Reinhold Publishing Corp.

Brown, L.R. 1970. Seed of Change-The Green Revolution and Development in the 1970's. Pall Mall Press, London.

Donald, C.M. 1968. The breeding crop ideotypes. Euphytica 17: 385-403.

Frankel, O. H. 1968. Third International Wheat Genetics Symp., Canberra, Australia.

RBI 2013. Handbook of Indian Economy. Reserve Bank of India, Mumbai

Simmonds, N. W. 1979. *Principles of Crop Improvement*. Longman, London and New York.

Singh, B. D. 2004. Plant Breeding: *Principles and Methods*. Kalyani Publishers, New Delhi.

Sneep, J. and Hendrikesen, A.J.T. (eds.) 1979. *Plant Breeding Perspectives*. Pudoc, Wageningen, Netherlands.

Genetic Basis of Plant Breeding

1. GENETIC CONSEQUENCES OF HYBRIDISATION

The basic theme of plant breeding is the management of genetic variability to develop superior varieties. Genetic variability stems from mutation, hybridisation and segregation. Genes who are the ultimate units of segregation follow certain rules of inheritance and behaviour after hybridisation. These rules of inheritance and behaviour in a population are simpler for major genes than for polygenes. The main reasons for this are that many polygenes govern the same character and are influenced by environment to a great deal. Unfortunately, most characters of economic importance are governed by polygenes. Therefore, following hybridisation a series of complex events are initiated while manipulating these characters for breeding varieties. The genetic basis of these characters and behaviour of these genes in a population have to be understood properly. Only after that populations can be created and handled by appropriate breeding method to manipulate gene and make effective selection in a crop species or in a particular situation.

1.1 Expression of Genes

The gene determines expression of characters, may be morphologic, anatomic, agronomic, physiologic or biochemical. The appearance of the first hybrid generation depends on the type of interaction between genes coming from the parents. If the expression of the character depends only on the interaction between allele at the same locus, it is simply a matter of dominance and recessive relationship. If the expression depends on interaction between genes at different loci, it is epistasis or non-allelic interaction.

Normally a gene expresses itself fully but it depends on penetrance and expressivity. Penetrance can be defined as the ability of a gene to be expressed in individuals, which carry it. Expressivity denotes the manner in which a gene is expressed. Thus, penetrance and expressivity denote the extent and manner of expression, respectively. For example, the bean variety Venutra carries a dominant gene that causes the tips and margins of unifoliate leaves to be partially chlorophyll deficient in 10 per cent plants of the population. The other 90 per cent, even though having the same genotype, do not exhibit the symptoms. In these 10 per cent plants, some have total deficiency and others have it at tips or margins. Thus, it may be said that this gene has 10 per cent (incomplete) penetrance and variable expressivity.

1.2 Segregation of Genes

Genes in hybrid re-combine and segregate in F_2 generation following the Mendelian Laws of inheritance. Thus, Mendelian segregation pattern for one, two and three genes would be in the ratio of 3:1, 9:3:3:1 and 27:9:9:9:3:3:3:1, respectively. With more number of genes, interallelic interactions, linkage, etc, the situation would be still more complex. This has been shown in Table 2.1.

Table 2.1   Numerical characteristics of hybrids between parents differing in n allelic pairs

No. of allelic pairs	Kinds* of gametes in F_1	Kinds of genotypes in F_2	Minimum size of F_2 where all genotypes can occur
1	2	3	4
2	4	9	16
3	8	27	64
10	1024	59,049	10,84,576
n	2^n	3^n	4^n

*Number of homozygous genotypes in F_2 would also be the same. Thus if two parents differ by n number of genes, the kind of gametes produced by F_1 plant would be 2^n. The number of homogygous genotypes would also be 2^n. In F_2, there would be different 3^n genotypes. The smallest F_2 population where these segregants can be recovered would be 4^n.

1.3 Linkage

Linkage is the tendency of parental genes to move together due to their location in proximity on the same chromosome. Linkage increases the frequency of parental types and reduces the frequency of recombinant types compared to that expected with independent segregation. There are two linkage phases: the coupling phase, where both dominant or both recessive genes come from the same parent, for example, AABB and aabb parents; and the repulsion phase, from where the dominant genes come from different parents, for example, AAbb and aaBB parents. This inequality depends on the intensity of linkage expressed as recombinant type is 0.5, the assortment is independent i.e. there is no linkage. In Table 2.2, the effect of different intensities of linkage on the recovery of parental and recombinant types is shown.

Table 2.2   Intensity of linkage and the recovery of AABB segregants in F_2

Recombination value (0.50 = no linkage)	% of AABB individuals in F_2 if the F_1 is	
	AB / ab (coupling)	Ab / aB (repulsion)
0.50	6.25	6.25
0.25	14.06	1.56
0.02	24.01	0.01
P	$\frac{1}{4}(1-P)^2$	$\frac{1}{4}P^2$

Linkage may be a boon to plant breeder if favourable or desirable genes are linked. For example, Rio and Turkey genes for bunt resistance in wheat are linked. In segregating population, selecting plants carrying Rio ensures 91 per cent of bunt resistance of Turkey gene also. If unfavourable genes are linked, a large population or special techniques would be required to break the linkage. For example, resistance to crown rust in oats is closely linked with susceptibility to Victoria blight. To breed a variety resistant to both diseases was difficult till a resistant source was induced by mutating the susceptibility loci for blight (see Chapter 12).

1.4   Speed of Homozygosis

Homozygosis is attained following inbreeding or selfing. Mendel himself had shown that starting with a heterozygote Aa, continued selfing decreases heterozygosity by 50 per cent in each generation. Thus, in F_1, the heterozygosity is 100 per cent, i.e., all the plants are heterozygous. In F_2, the heterozygosity is reduced to 50 percent. For example, Aa hybrid would give AA, Aa, aa segregants in 1:2:1 proportion and half of it would be homozygous. Any number of heterozygous gene pairs would follow the same rule. Thus, with n heterozygous gene pairs, the proportion of completely homozygous plants after m generation of self fertilisation equals $\{(2^m - 1)/2^m\}n$. The per cent of homozygosis met within different generations of selfing when 1,5,20, and 100 gene pairs segregate, as obtained by the expansion of the above formula, is depicted in figure 2.1.

2.   QUANTITATIVE INHERITANCE

In the classical work of Mendel, the parents for crossing had contrasting characters and to his luck, each character was governed by one gene. In the F_2 generation the segregation was in discrete classes of red and white or tall and dwarf. Thus, there was no doubt as to whether one of his plants was tall or dwarf or its flowers red or white. Such inheritance is known as qualitative inheritance and the characters are called qualitative characters.

In contrast to these, there are a number of characters such as yield, maturity, grain dimension, etc, which can be specified only by using metrics such as weight, time, length, etc these are called metric or quantitative characters. In their inheritance, segregation in discrete classes is not observed in F_2. For example, if a cross is made between low and high yielding varieties, the segregants in F_2 cannot be fitted into any discrete ratios, say 3:1;9:3:3:1 or any modifications thereof. The segregants would show a continuous range from low to high yielding (Figure 2.2) types. Clearly in such cases the range

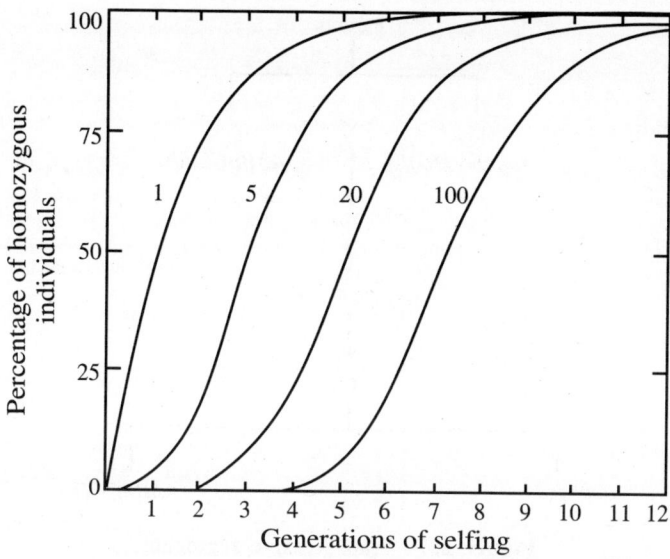

Figure 2.1 Number of genes involved and speed of homozygosis. Values 1,5,20,100 indicate number of genes involved.

of F_2 variability can be divided into classes with frequency of segregants in each (See Figure 2.2). such F_2 distribution is characterised by certain statistics like means, variances and covariances. Fisher (1918) laid the foundation of biometrical genetics for interpreting the above statistics in Mendelian sense.

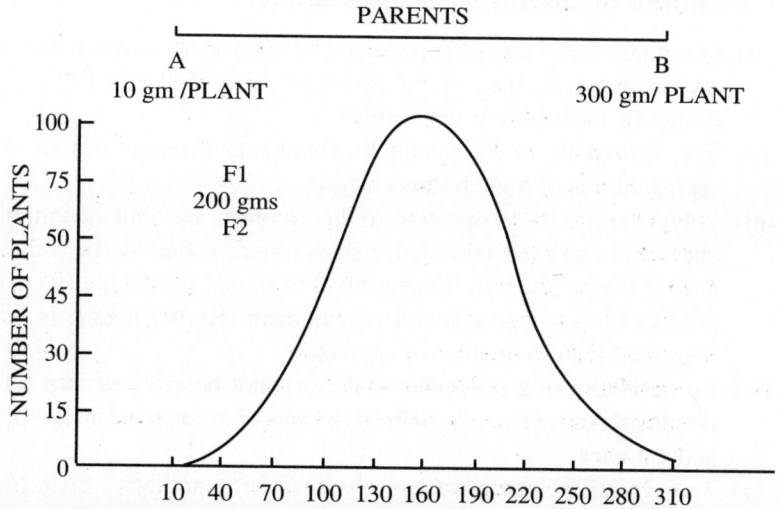

Figure 2.2 Chief features of quantitative inheritance indicating in F_2 generation a continuous pattern of segregants are obtained.

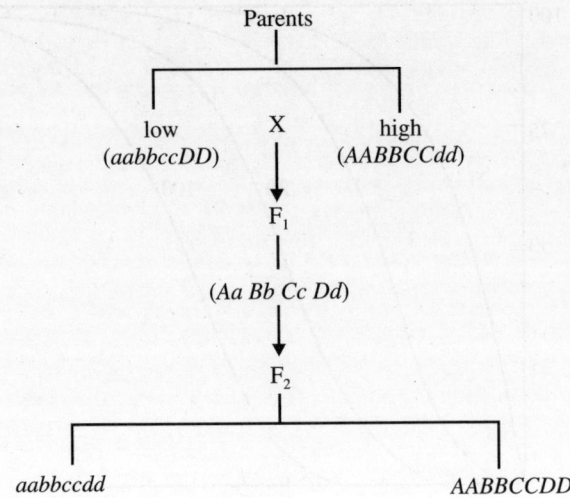

Figure 2.3 Transgressive segregant.

Around that time Nilsson Ehle (1909) established the basis for quantitative genetics by his discovery of additive effects of 3 loci affecting grain colour in wheat. It was also in 1909 when Johannsen established that heritable and non-heritable (environmental) variations are jointly responsible for continuous variations. The inheritance pattern of such variations is assumed to be polygenic.

2.1   Features of Quantitative Inheritance:

(i)   Quantitative characters are governed by many genes called polygenes each with small effect on the genotype. Individual effect of a gene cannot be measured or determined.

(ii)   The segregants in F_2 cannot be fitted into discrete classes. The segregation is of a continuous nature.

(iii)   Polygenes are more sensitive to environment and thus quantitative characters and their inheritance shows more variation. The relation $P = G \times E$ was given by Johannsen (1903), that phenotype (P) is the product of genotype (G) and environment (E). but here E is more important than in qualitative characters.

(iv) F_2 variability in a polygenic system cannot be grouped into a few discrete classes; rather it is defined in terms of mean, standard deviation and variance.

(v)   Transgressive segregants are observed in quantitative characters. Transgressive segregation refers to the appearance, in F_2 or later generations, of phenotypes, which transgress the maximal and minimal character expressions of the two original parents. Transgressive segregation can only arise when one or both of the parents do not represent the extreme genotype. For example, if low and high yielding parents in the above example do not represent the extremes, a transgressive segregant may arise in the following manner.

Thus the segregants in F_2 with lower yield (aabbccdd) than low yielding parent and with higher yield (AABBCCDD) than high yielding parent are possible to be recovered.

2.2 Population Structure

Population in genetic sense is a community of individuals, which share a common gene pool. Thus individuals of a population are related ancestrally. Population of a self pollinated crop would be heterogeneous and consist of homozygous individuals. Mutation is the only source of genetic variation in self pollinated crops. The mutations so created would also become established in the population as homozygous individuals under self pollination. Therefore the population would carry homozygous individuals of various genotypes and would be called as heterogeneous and homozygous. It the population consists of the progenies of only one homozygous individual; it would be called homogeneous and homozygous.

Contrary to the above, in cross pollinated crops each individual plant would be heterozygous and a natural population thus would be heterogeneous and heterozygous. A population where each individual has equal chance of mating with other, that is, can intercross freely, is called a random mating population or Mendelian population.

2.3 Hardy-Weinberg Law

Hardy (1908) in England and Weinberg (1909) in Germany formulated the law independently, which is known as Hardy-Weinberg Law. This law states that in a random mating population the gene frequencies of a given allele remains constant generation after generation unless altered by selection, non-random mating, differential migration and mutation. For example, if A and a alleles are in q and 1-q frequency, then this would remain constant in the genotypic frequency : q2AA+2q(1-q) Aa+(1-q)2aa=1

To understand more fully the basic principles governing the frequencies of alleles, one might consider a population derived over a number of generations from intercrossing two heterozygous plants Aa x Aa. The frequency of two alleles is equal, that is, A = a = 50% = 0.5. In the first generation AA, Aa and aa would appear with the genotypic frequency of AA = 0.25, Aa = 0.5 and aa = 0.25. Then in the subsequent generation under random mating:

♀♂	A (0.50)	a (0.50)
A (0.50)	AA (0.25)	Aa (0.25)
a (0.50)	Aa (0.25)	aa (0.25)

Genotypic frequency of AA = 0.25
AA = 0.25
Aa = 0.25 + 0.25 = 0.50

Gene frequency
A = 0.25 = 0.25 = 0.50
a = 0.25 + 0.25 = 0.50

Figure 2.4   Illustration and explanation of gene frequency remaining unchanged subsequently

This shows that gene frequencies have remained unchanged. This would remain unchanged as long as the differential selection, migration, mutation from A-a or a-A and non random-mating does not take place.

3. COMBINING ABILITY

This is the ability of a parent to produce inferior or superior combinations in one or a series of crosses. Combining ability is of two types:

3.1 General Combining Ability

General combining ability (g.c.a.) is the average performance of parent in a series of crosses. For example, if parent A is crossed to 10 different parents, then its g.c.a. would be the average performance of its 10 hybrids.

3.2 Specific Combining Ability

Specific combining ability (s.c.a.) is the combining ability of a parent in a specific cross. This is estimated as a deviation in a particular cross from performance predicted on the basis of g.c.a. for example, the 10 hybrids of A in the above crosses may not be uniformly good. Some may be extremely heterotic and superior. In that case that s.c.a. of the parent in that particular cross would be rated as high. It is usually expressed as the deviation in the performance of a specific cross from the performance expected on the basis of g.c.a. of the parent. Depending on the pattern of crosses under study, g.c.a. and s.c.a. are calculated using various methods.

4. HERITABILITY

Johannsen (1903) demonstrated that phenotype (P) is the interaction product of genotype (G) and the environment (E). thus a plant selected on the basis of phenotype may not perform the same in the next generation with the slightest change in the environment. The degree of resemblance of original and selected plant would depend on the degree of role of G and E in determining P. If E plays a more important role, the plant would not perform closer to the original one. It may be said then that the character under reference has low heritability. Vp is the phenotypic variance, Vg can be further divided into D and H which are the additive and dominant components of genetic variance, respectively. Heritability is the ratio of genetic (Vg) and phenotypic (Vp) variances, even though written as h^2.

If the genotype completely determines the phenotype, both the variances would be equal and heritability would be 100 per cent.

Since,
$$Vp = Vg + Ve$$
$$Vg = Vp - Ve$$

Heritability (h^2) can be calculated, in broad sense, on the basis of the following formula:

$$h^2(\%) = \frac{Vg}{Vp} \times 100$$

There are several methods to estimate Vp, Vg and Ve, but the one commonly used is as follows: Two parents with contrasting differences, say low and high yielding, are crossed to produce F_1 and F_2 generations and back crossed with both parents to produce B_1 and B_2 generations. A field trial using parents (P_1, P_2), F_1, F_2, B_1, and B_2 is conducted and observations on yield of single plants are recorded. Variance is calculated for each generation, namely variance for F_2 (VF_2), variance for back cross generations (VB_1, VB_2) and variance of parents (VP_1, VP_2). The following is the genetic expectations for various variances:

(1) $VF_2 = 1/2\ D + ¼\ H + E$ = phenotypic variance = Vp

(2) $VB_1 + VB_2 = ½\ D + ½\ H + 2E$

(3) $\dfrac{(VP_1 + VP_2 + VF_1)}{3} = V_e$

Where D is the additive variance, H is the variance due to dominance deviation and E is the environmental variance. Since any variation from plant to plant in parental or F1 population is solely environmental, the average of 3 non segregating generations (P1, P2 and F_1) estimates environmental variance (Ve).

$2VF_2 - (VB_1 + VB_2) = \frac{1}{2} D$ = additive genetic variance

$VF_2 - Ve = Vg = \frac{1}{2} D + \frac{1}{4} H$ = total genetic variance

Heritability (h^2) may be calculated by two ways:

(i) h^2 (%) = $\frac{1}{2} D / VF_2 \times 100$ (heritability in a narrow sense)

(ii) h^2 (%) = $Vg / Vp \times 100 = \frac{1}{2} D + \frac{1}{4} H) / Vp \times 100$ (heritability in a broad sense)

Since only additive genetic variability is inherited in the next generation, heritability in a narrow sense is more reliable and is used in breeding.

5. GENETIC ADVANCE

Selection in breeding programme are made to improve a population or to create a superior type. The advance made by selecting desired types is known as genetic advance under selection. This is equal to the difference in the mean of two populations.

$$\text{Genetic Advance} = k \times \sqrt{Vp} \times h^2$$

where k , $\sqrt{Vp}$ and h^2 denote selection differential, phenotypic standard deviation, heritability coefficient respectively. Usually, the best 1, 2 or 5 per cent plants are selected from the original population. In that case the value of k would be 2.64, 2.42 and 2.06, respectively. Phenotypic standard deviation is calculated on the original population, which simply measures the amount of variability present in that population. Clearly then, if heritability is high, variability is more and the top 1 per cent plants are selected, it will give maximum genetic advance.

6. GENETIC BASE

Human race has utilised 3,000 species of plants for food, of which 150 have entered world commerce but the world population is fed by only 16 crops, namely, rice, wheat, maize, sorghum, bajra, barley, sugar cane, sugarbeet, potato, sweet potato, cassava, common bean, soybean, groundnut, coconut and banana. In these crops also a few cultivars are predominant, as in wheat Sonalika. Where there are a number of varieties, there are also some gene in common to all like DGWG dwarfing gene and *Cina* cytoplasm in most of the rice varieties, *milo* male sterility cytoplasm in all hybrid sorghum varieties, Tifton male sterility cytoplasm in all bajra hybrids, and *Tms* cytoplasm in all sterility based maize hybrids. Only two genes, *Rht* 1 and *Rht* 2, from Norin 10 donor have been used in all dwarf varieties of wheat of the world (Gale and Marshall, 1979). Even extreme are the entire coffee plantations in South America, which originated in 1706 from one single coffee plant grown in Amsterdam Botanic Garden, Netherlands.

Thus the base of these crop species is narrow as well as the cultivars have genetic uniformity and hence form an extreme kind of mono culture. Any outbreak of disease on one crop variety would attack the other. For example, the southern corn blight due to T race of *Helminthosporium maydis* attacked all maize hybrids having *Tms* cytoplasm. The area under hybrids carrying Tms cytoplasm was above 85 per cent in 1970 and so a disastrous crop loss took place in the USA.

Therefore it is recommended that the monoculture varieties should be avoided and instead varieties with diverse genetic background should be used (Anon. 1972). In breeding, alternative sources of dwarfism, as in wheat Rht 3 gene from Tom Thumb, should be used.

7. CHOICE OF BREEDING METHODS

The majority of commercially important plant characteristics is quantitative and usually displays continuous variation. Plant population are therefore usually characterised by statistical quantities which measure average expression (means) and the range of expression (variances). These quantities are partitioned by biometrical methods into genetic components that can be utilised in designing a breeding procedure.

Breeding methods are dictated by the action, interaction and linkage relationships of the genes controlling the characters. These also decide about the size of population to be handled and the method of promoting and advancing the segregating generations.

By knowing the relative magnitude of additive, dominant and ecpistatic components of variation, proper breeding method can be designed. On realisation, for example, that additive gene actions are important, synthetics and composites were bred in place of hybrids in cross pollinated crops. There is even a powerful argument that the production of hybrid varieties should on genetic grounds be complementary to, not exclusive of, the production of homozygous varieties.

8. USEFUL REFERENCES

Allard, R.W. 1960. *Principles of Plant Breeding*. John Wiley & sons Inc., New York

Anon 1972. Genetic Vulnerability of Major Crops, NAS, Washington.

Brewbaker, J.L. 1964. *Agricultural Genetics.* Prentice-Hall, Inc., New Jersey

Chopra, V. L. (Ed.) 2000. *Plant Breeding: Theory and Practice.* Oxford & IBH, New Delhi.

Falconer, D.S. 1960. *Introduction Quantitative Genetics.* Oliver Boyd, Edinburgh

Gale, M.D. and Marshall, G. A. 1979. Classification of Norin 10 and Tom Thumb dwarfing genes in hexaploid bread wheat. Proc. V. Int. Wheat Genetics Symp. 1978. New Delhi, 995-1001.

Hutchinson, J.B. 1965. Essay on Crop Plant Evolution. Cambridge Univ. Press, London

Lerner, I. M. 1958. *The Genetic Basis of Selection.* John Wiley & sons Inc., New York

Mayo, O. 1980. *The Theory of Plant Breeding.* Clarendon Press, Oxford

Mather, k. 1967. *The Elements of Biometry.* Methuen, London

Mather, K. and Jinks, J.L. 1977. *Introduction to Biometrical Genetics.* Chapman and Hall Ltd., London

Reeve, E.C.R. and Waddington. C.H. 1952. Quantitative Inheritance. HMSO, London

Williams, W. 1964. *Genetical Principles and Plant Breeding.* Blackwell, Oxford

Nature of Crops and Methods of Breeding

The breeding methods applicable to a crop species depend, to a great extent, on its mode of reproduction and flower morphology. Therefore, these aspects have to be known earlier for the choice of breeding methods. A brief review on these aspects is presented below keeping only crop plants in mind.

1. MODE OF REPRODUCTION

There are three modes of reproduction in crop plants: vegetative, sexual and apomictic.

1.1 Vegetative

Also known as asexual reproduction, this mode is prevalent in all those plants devoid of seed set, long reproduction cycle, heterozygosity, etc. Various plant parts, like normal or modified stems, roots, leaves etc, known as propagule, are used for multiplication.

(i) **Normal Stem:** Normal stem is used as a propagule by cuttings (e.g., sugarcane, sweet potato), airlayering or gootie in lemon, oranges and other citrus, layering (e.g., lemon, litchi), grafts e.g., mango, apple).

(ii) **Modified Stem:** Modified stems like rhizome (e.g., ginger, turmeric, colocasia), tuber (e.g., potato), bulb (e.g., onion, garlic), corm (e.g., Colocasia, yam), stolon (e.g., strawberry), sucker (e.g., banana, Chrysanthemum, Mentha) are used as propagation material.

(iii) **Normal and modified Root:** Normal roots of wood apple and citrus and many such trees are used for propagation, as these roots have potential to form bud and stem. Modified roots such as tuberous (sweet potato), fasciculated roots (dahlia, asparagus), etc, are also used as propagule.

(iv) **Others:** Many other parts like bud for budding in roses, bulbils (modified flower) in some garden plants, leaf, etc, are also used as propagule to grow and multiply a crop.

1.2 Apomictic

Apomixis is a mode of reproduction where sexual organs of flowers propagate in a vegetative manner. Apomictic are the plants that have apomixis mode of reproduction. Apomixis may be defined as the substitution for sexual reproduction of an asexual process, which does not involve any nuclear fusion. Apomixis represents a transition phase between sexual and asexual modes of reproduction. It has been reported in more than 300 genera of 80 families. Maheshwari (1950) classified apomixis into following four groups:

(i) **Non-recurrent:** The megaspore mother cell undergoes the usual meiotic division to form the haploid embryo sac. The embryo may arise either from the egg (haploid parthenogenesis) or from some other cell of gemetophyte (haploid apogamy). The plants resulting from this type of apomixis are always haploid. Thus, haploids reported in crops like maize, rice, tomato, etc, originate through non-recurrent apomixis.

(ii) **Recurrent:** The embryo sac may arise either from a cell of the archesporium (generative apospory) or from some other part of nucellus (somatic apospory). There is no reduction division. The embryo may arise either from the egg (diploid parthenogenesis) or from any other cell of gametophyte (diploid apogamy). The resulting plants from recurrent apomixis are diploid. Hybrid (heterozygous) plants with recurrent apomixis (parthenogenesis) breed true and behave like a "permanent hybrid".

(iii) **Adventive embryony:**   When the embryo develops from cells of nucellus or integuments but not from gametophyte, it is called adventive embryony or sporophytic budding or polyembryony. Since the unreduced ovule functions as diploid cell, it is just like vegetative propagation and there is no alternation of generation. Adventive embryos have been reported in citrus, mango, *jamun* (*Syzygium cumini*). Usually on germination such seeds give more than one seedling-as many as 13 in citrus- but all are diploid.

(iv) **Vegetative apomixis:**   In this case seeds in flowers are replaced by structures like bulbils, etc, as in *Poa bulbosa* and onions.

1.3   Sexual (Amphimixis):

This type of reproduction involves the fusion of gametes produced by male (pollen) and female (ovary), and formation of seed. Since male and female sexes are involved, it is known as sexual reproduction or amphimixis. Depending on the nature of pollination, this group is further divided into four categories.

(i) **Naturally or normally self-pollinated:**   Those crops, which are predominantly self-pollinated and cross-pollination is less than five per cent come in this category. For example, wheat, rice, barley, ragi (minor millet), oats, beans, cowpea, pea, chickpea, urd, mung, groundnut, sesame, soybean, tobacco, jute and tomato fall in this group.

(ii) **Often self-pollinated crops:**   Those crop which are self-pollinated but the extent of cross pollination is more than five per cent. For example linseed, yellow sarson, pigeon pea.

(iii) **Naturally or normally cross-pollinated:**   These are cross-pollinated group of plants but self-pollination may occur up to five per cent. Some of these crops are alfalfa, castor, hops, maize, rye, apple, almond, sugarbeet, cabbage, cauliflower, most cucurbits, carrot, coriander, cherries, citrus, palm, fig, grape, melons, mango, onion, papaya, radish, spinach and strawberry.

(iv) **Often cross-pollinated:**   These crops are by nature cross-pollinated but more than five per cent self-pollination occurs. For example, cotton, sorghum, jute, tobacco etc.

2.   INCOMPATIBILITY

The term incompatibility is used to describe the failure of pollen to germinate, the pollen tubes to penetrate the full length of the style or to fertilise the egg even when pollen and ovule are functional. Incompatibility differs from sterility as in the latter the gametes (pollen or ovule) are non-functional. Self-incompatibility prevents self-fertilisation and thus promotes cross pollination. But it is a superior mechanism for cross-pollination than dioecy or monoecy where a large amount of pollen is produced unnecessarily and also certain ovules get self-fertilised. It has been reported in more than 3000 species of plants covering important families such as Leguminoseae, Rosaceae, Solanaceae, Compositae, Cruciferae and Gramineae.

Self-incompatibility is of two types: homomorphic and heteromorphic.

2.1 Homomorphic System

In these, there is only one type of flower, that is, compatible and incompatible plants cannot be differentiated on the basis of flower morphology. There are two types of incompatibility systems in these species.

(i) **Gametophytic incompatibility:** This was first reported by East and Mangelsdorf (1925) in *Nicotiana sanderae*. Gametophytic incompatibility is governed by multiple alleles of S gene in pollen, stylar tissue and ovule. As many as 212 multiple alleles have been found in *Trifolium* species. These genes are independent in their action and do not show any dominance relationship.

Pollen and ovule carrying dissimilar S alleles can unite to set seed if the style does not inhibit the tube-growth. The stylar tissue is diploid and if there is any S allele in common with pollen, it arrests the tube growth. For example, styles of S_1S_2 genotype would not allow any pollen tube carrying either S_1 or S_2 allele to penetrate whereas it will not inhibit S_3 and S_4 pollen tubes. Thus there are three situations: fully incompatible, partially incompatible, and fully compatible (Figure 3.1). Gametophytic incompatibility has been reported in tomato, pear, peaches, mango, tobacco, lucernes, rye, etc.

(ii) **Sporophytic incompatibility:** This was first described by Hughes and Babcock (1950) in Crepis and by Gerstel (1950) in *Parthenium*. This system is governed by multiple alleles of S gene like the gametophytic system but genes are not independent and show dominance relationship. The pollen tube reaction is not governed by its own genotype but by the genotype of the plant on which it is produced. For example, both types of pollen produced by S_1S_2 plant would behave similarly. The other difference with gametophytic system is that one allele would be dominant on another. Thus, if S_1 is dominant, S_1S_2 style would behave like S_1S_1 style.

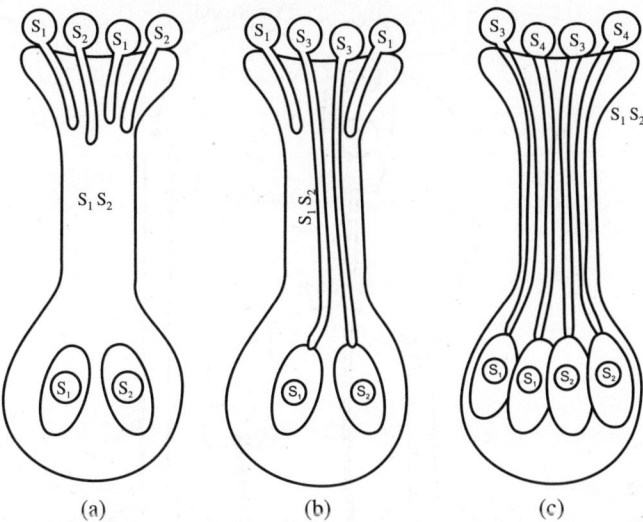

Figure 3.1   Gametophytic incompatibility: S_1 to S_4 are incompatibility alleles. a : is fully incompatible; b : half incompatible; c : fully compatible

Inhibition occurs at the stigmatic surface with the result that the pollen does not germinate on stigma. Sporophytic incompatibility has been reported in *Brassica oleracea, B. campestris, Cosmos*, etc.

2.2   Heteromorphic System

These species have two or three types of morphologically distinct flowers. Each type is self incompatible. At the incompatibility locus there are two alleles closely linked with genes affecting style and filament length.

(i) **Distyly:**   In Primula there are two types of flower, 'Pin type' with long style, short filament, large stigmatic cells and small pollen and 'Thrum type' with short style, long filament, small stigmatic cells and large pollen (Figure 3.2). both types are self incompatible but cross compatible. The thrum is governed by Ss genes and pin by ss. The pollen of thrum plant, in spite of their segregation into S and s types, behave as if all of them had S genotype. This is because the reaction of pollen is controlled by the genotype of the sporophyte. Distyly is also found in *Linum*.

(iii) **Tristyly:**   In Lythrum, there are three types of flowers (Figure 3.3) long, mid long and short. Each type has two positions of anthers. The genetic constitutions of the three forms are as follows:

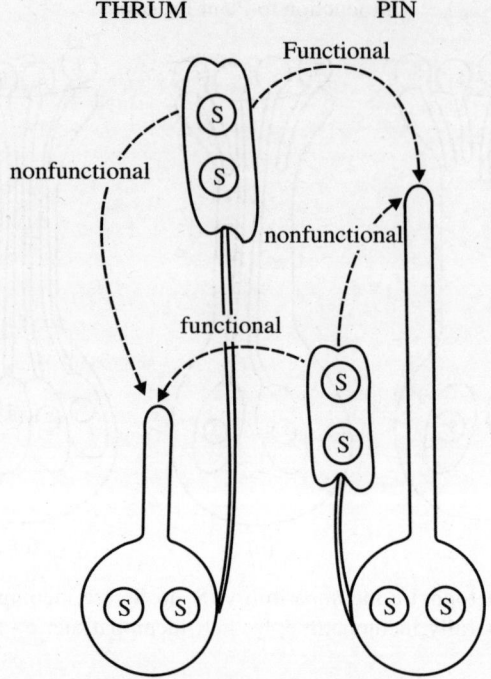

Figure 3.2   Distyly in *Primula*. Pin and Thrum types of flowers are self incompatible but cross compatible.

Long style:	mmss
Mid style:	Mmss or MMss
Short style:	MmSs, mmSs, mmSS, MMSS, MMSs, MmSS

Two independent loci S and M govern the style length. Plants with S have short style irrespective of their M constitution. Plants with s have medium styles if they have M and long style if they have m. All the three stylar types are self incompatible but cross compatible (Figure 3.3).

3.   USE OF INCOMPATIBILITY IN PLANT BREEDING

Incompatibility may be used in hybrid seed production without emasculation as self fertilisation will not take place. Leffel (1963) suggested the production of double cross hybrids (S1S1 × S2S2) × (S3S3 × S4S4) in red clover utilising four incompatibility alleles. Thompson (1964) suggested the use of six S genes for the production of double cross hybrid kale (see Chapter 9 and figure

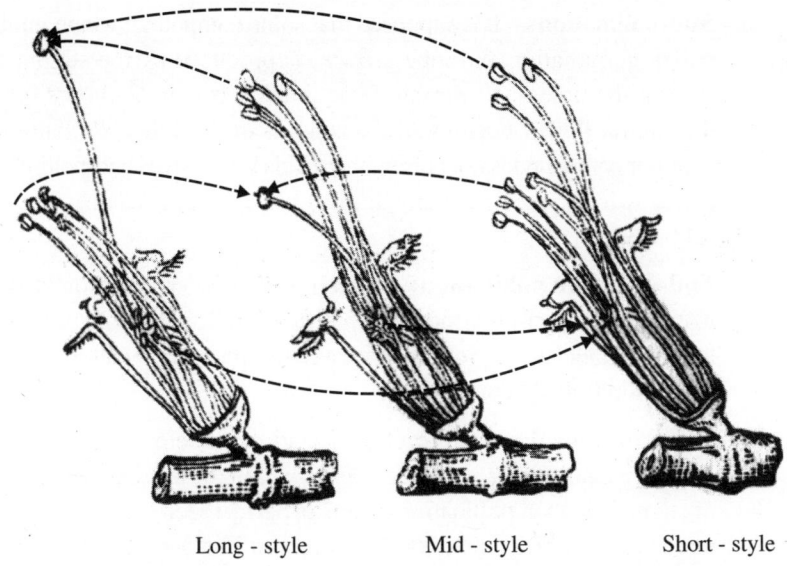

Long - style   Mid - style   Short - style

Figure 3.3   Tristyly in *Lythrum* (after Williams, 1964). Short styled flowers
are compatible to medium and long; long styled flowers to short
and medium; and medium to long and short styled.

9.3). Individual lines may be maintained by selfing either in bud (bud pollination)
or late stage.

Fruit breeders are now conscious of incompatibility system and avoid
planting pure orchards of mango, pear and peaches. Two or more compatible
varieties are inter-planted. But it has been reported that self compatible varieties
are more successful in cultivation; thus efforts are made to breed self compatible
varieties. In cherry, self compatible varieties have been bred by X-ray treatment
of incompatible varieties.

3.1   Pseudo-incompatibility

In incompatible plants some degree of self-fertilisation may be affected. This
is known as pseudo-compatibility. These avenues are used for creating
homozygous S lines and maintaining them by selfing.

3.2   True-incompatibility

In true-incompatible plants no self-fertilisation takes place under natural
conditions. However, the following ways are used to maintain the self
incompatible species and strains by self pollination.

(i) **Bud-pollination:**   It is supposed that some compound, which inhibits pollen germination and tube growth, is produced by the stigma and style at the time of flowering. If pollination is done 24 hours before the natural flower opening, there is no incompatibility. Probably the inhibitor compound is not present then. This is known as bud-pollination, and is widely used in maintaining lines in rapes, cauliflower and cabbage, etc.

(ii) **End-of-season pollination:**   Usually when the crop season is about to end, incompatibility mechanisms weaken. This is probably due to less metabolic activity of the plant and failure of the plant to produce enough inhibitory compounds.

(iii) **Stigma removal:**   Failure of pollen to germinate or tube to grow occurs in cauliflower, cabbage and radish. In these cases the removal of stigma and then pollination results in perfect seed set. This suggests that incompatibility reaction develops on contact of pollen grains with stigmatic layers.

(iv) **Grafting:**   In *Trifolium partense*, grafting of vegetative parts, either within a single plant (homografts) or between different plants (heterografts), have also improved the level of compatibility as reported by Evans (1959). The reason is not known. Pollination can be done after grafting.

(v) **Making polyploid:**   If two diploid incompatible strains are made tetraploid, they may become compatible because of the altered genic balance at the tetraploid level. The method has been utilised in pear and potato species. The reason is shown in figure 3.4.

4. MALE STERILITY

Male sterility can be defined as the incapacity of plants to produce or to release functional pollen. In some cases pollen grains are not formed. These are called pollen free type. In others, pollen grains abort during the process of development. These are called pollen aborted or pollen sterile type. In still others, viable pollens are formed but anthers fail to dehisce. This is called functional male sterility. Since male sterile plants can be used as female parent without emasculation, they have been used in hybrid seed production. Based on the patterns of inheritance, there are three types of male sterility.

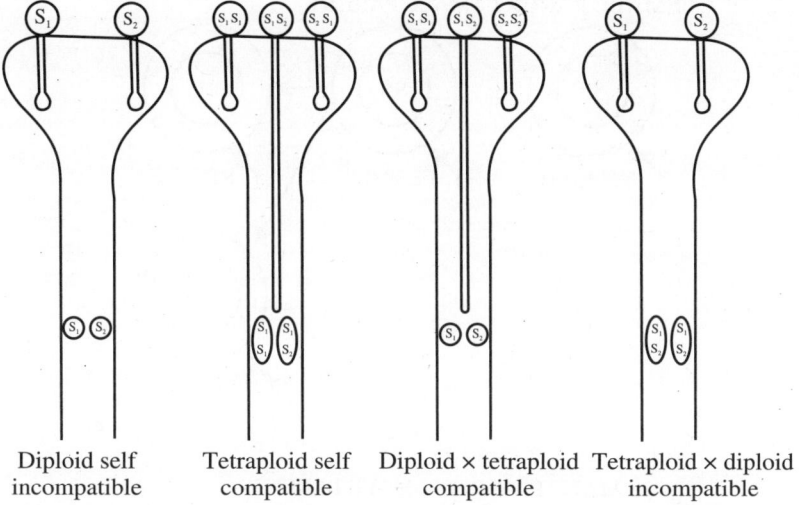

| Diploid self incompatible | Tetraploid self compatible | Diploid × tetraploid compatible | Tetraploid × diploid incompatible |

Figure 3.4   Breaking incompatibility using poyploidy technique

4.1   Genetic Male Sterility

This type of sterility is dependent on the action of genes carried in the nucleus. Usually this is governed by recessive (ms) genes. The inheritance pattern is shown in Figure 3.5. Genetic male sterility has been reported in tomato, barley, brinjal, rice, and soybean. This type of male sterility is maintained by crossing male sterile plants with fertile ones and recovering the male sterile segregants (1 sterile : 1 fertile). Genetic male sterility has not been used much in practical plant breeding owing to problems in its maintenance. But it is used in creating a open pollinated population in self pollinated crops by introducing male sterile gene in it. By selecting seed set only on male sterile plants, breeders select only hybrid seed and later can recover unique genetic combinations. This method is called genetic male sterility facilitated breeding.

Expression of genetic male sterility (*ms*) may depend on the temperature or photoperiod (light duration) a plant receives. If the action of *ms* gene is dependent on the temperature it is called Temperature Sensitive Genetic Male Sterility (TGMS). In this case the same line (variety) becomes fully male fertile or fully male sterile depending on the temperature. If the action of *ms* gene is dependent on the duration of light the plant receives, it is called Photoperiod Sensitive Genetic Male Sterility (PGMS). By identifying suitable places of temperature or day length range, the same line (variety) can be used as male sterile or male fertile. It saves time and money in commercial seed production.

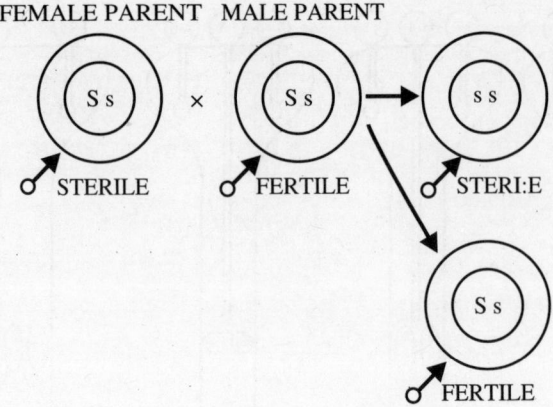

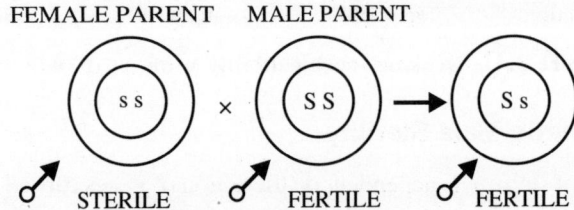

Figure 3.5   Genetic male sterility and its inheritance. The gene S is dominant over recessive gene s.

Male gamete is more sensitive than female and hence any variation in the growing environment makes plant male sterile. Other than the environment, chemicals are also known to induce temporary (limited to the sprayed plant) male sterility. These chemicals are called male gametocide or chemical male sterilants. These chemicals belong to different groups like arsenics (zinc methyl arsenate, sodium methyl arsenate etc), ethrel, maleic hydrazide etc.

4.2   Cytoplasmic Male Sterility

This type of male sterility depends on cytoplasmic factors. Plants carrying particular types of cytoplasm are male sterile but will produce seed if pollinated by a male fertile plant. These F1 seeds produce only male sterile plants, however, since their cytoplasm is derived entirely from the female gamete. The maintenance of male sterile lines is easy (Figure 3.6) cytoplasmic male sterility has real advantages in ornamental species. This is because, offsprings of male steriles are always male sterile regardless of the pollen and remain fresh longer than their fertile counterparts. Cytoplasmic male sterility is also useful in producing single or double cross hybrids in crops where some vegetative part is the commercial product. In onion it has been utilised. It is obviously unsuitable for production of hybrid seed in crops where the fruit or seed is the commercial product. Cytoplasmic sterility has been well demonstrated in onion.

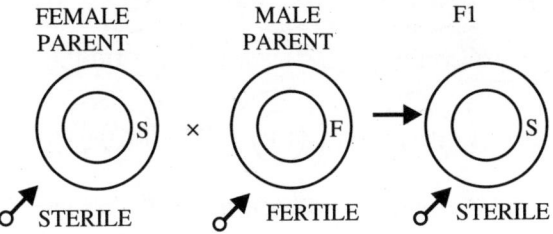

Figure 3.6   Cytoplasmic sterility and its inheritance. Letters S (Sterile) and F (Fertile) cytoplasmic factors

4.3 Cytoplasmic Genetic Male Sterility

As the name indicates, male sterility in this type is determined by the interaction of nuclear genes and cytoplasm. The inheritance pattern (Figure 3.7) show that a plant would be male sterile only when sterile cytoplasm (S) is present with rf nuclear gene. Plants with sterile (S) cytoplasm will, however, be fertile in presence of *Rf* gene. Plants with fertile (F) cytoplasm will be male fertile irrespective of *Rf* or *rf* gene. Thus, this sterility differs from cytoplasmic type in one respect, that is, some pollinators can restore its fertility. Cytoplasmic-genetic sterility has been reported in bajra, *Brassica napus*, carrots, chillies, maize, mustard, onion, pigeon pea, rice, sorghum, sunflower, tobacco, and wheat.

Duvick (1966) reported that in USA., 23 per cent of onion, 60 per cent of sugarbeet and 85 per cent of maize acreage was under hybrid seed produced by cytoplasmic-genetic male sterility. In India, almost the entire acreage under hybrid sorghum utilises cytoplasmic-genetic male sterile source CK 60 (Combined Kafir 60). Figure 3.8 depicts the production of sorghum hybrid. Similarly, hybrid bajra is produced by using cytoplasmic male sterile source MS Tift 23 A.

Cytoplasmic-Genetic-Male sterility (CMS) system is used extensively in producing hybrid rice in India, China, Vietnam, Myanmar, Bangladesh and a few more countries. Several million hectares of hybrid rice is produced in these countries currently.

Nature of Crops and Method of Breeding

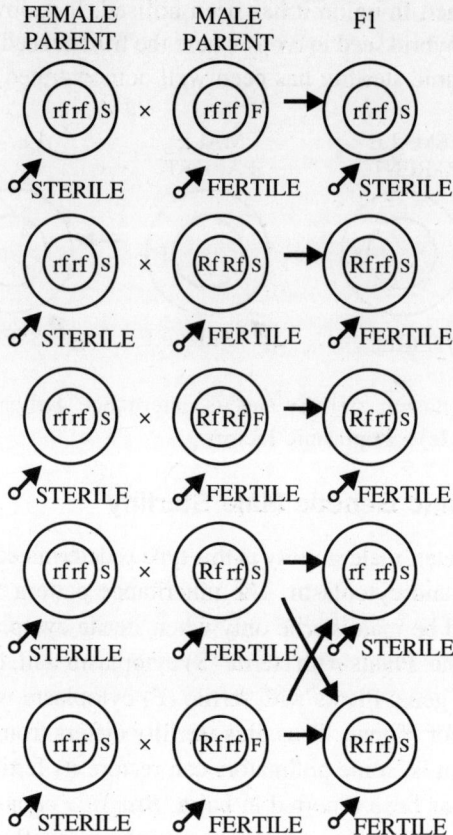

Figure 3.7 Cytoplasmic - genetic male sterility and its inheritance pattern. Letters in the "inner circle" represent genetic factors and in "outer circle" represent the cytoplasmic factor. Rf is the fertility restoring gene and is dominant over *rf*. Letters S and F denote the Sterile and Fertile cytoplasm respectively.

5. HYBRIDISATION

Hybridisation is the crossing of two genetically dissimilar individuals. The resulting seed is called a hybrid seed or F_1 seed and the plant coming out of F_1 seed is a hybrid plant or F_1 plant. Thus, *hybrid is a heterozygous individual for one or more genes.* Hybridisation as a method of plant breeding is more than 250 years old. Thomas Fairchild (1717) was the first to produce a plant hybrid by crossing two ornamental flower plants, Sweet William and Carnation. But one German botanist Kolreuter (1760) was first to use hybrids of tobacco

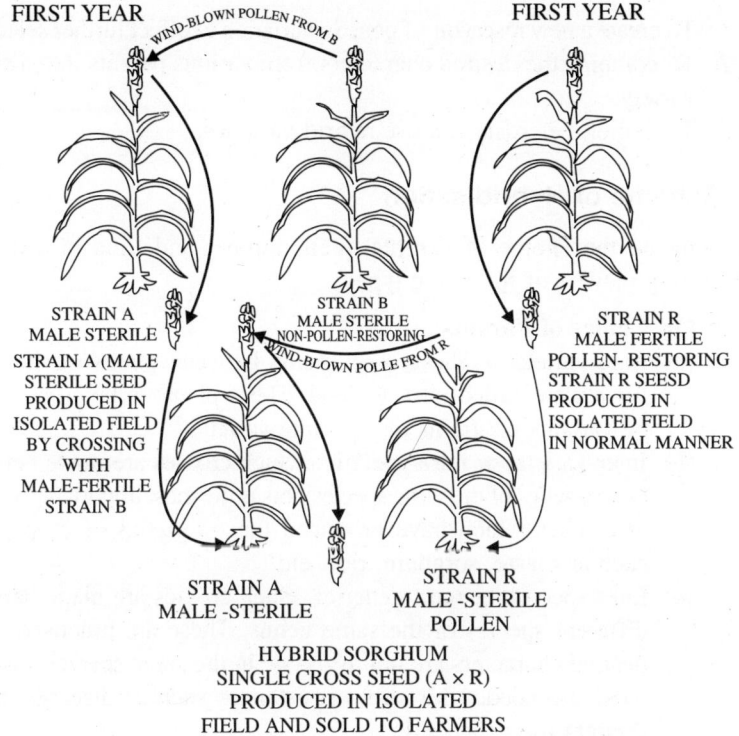

FIRST YEAR

WIND-BLOWN POLLEN FROM B

FIRST YEAR

STRAIN A
MALE STERILE

STRAIN A (MALE
STERILE SEED
PRODUCED IN
ISOLATED FIELD
BY CROSSING
WITH
MALE-FERTILE
STRAIN B

STRAIN B
MALE STERILE
NON-POLLEN-RESTORING

WIND-BLOWN POLLE FROM R

STRAIN R
MALE FERTILE
POLLEN- RESTORING
STRAIN R SEESD
PRODUCED IN
ISOLATED FIELD
IN NORMAL MANNER

STRAIN A
MALE -STERILE

STRAIN R
MALE -STERILE
POLLEN

HYBRID SORGHUM
SINGLE CROSS SEED (A × R)
PRODUCED IN ISOLATED
FIELD AND SOLD TO FARMERS

Figure 3.8   Use of cytoplasmic – genetic male sterility in producing sorghum hybrids.

for plant breeding. Many plant hybridisers including Luther Burbank, the charmer of botanic kingdom, made extensive inter-varietal and inter-generic crosses. But real insight into hybridisation as a method of developing superior varieties came only after the Mendelian laws of inheritance were rediscovered in 1900.

5.1   Objective

After the original genetic variability available from introduction and or natural sources has been exploited, a breeder is confronted with the need to create a new reservoir of genetic variation upon which to practice the selection. It is amply clear from previous chapter that once pure lines, inbreds or clones have been evolved, further selections within them are not effective. At this point, hybridisation among the selected types or mutation in them, are taken up to create new reservoirs of genetic variation upon, which further selections are made. Thus there are the following objects of hybridisation.

1. To create a new reservoir of genetic variation to effect further selection.
2. To combine the desired characters from various parents into a single variety.
3. To exploit heterosis and use hybrid varieties.

5.2 Scheme of Hybridisation

Depending on the choice of parents, their number and time of crossing, hybridisation may be of following types.

(i) **On Choice of Parents:**

   (a) Intra-varietal:– The crosses made between plants of the same variety are called intra-varietal. These plants to be crossed may be variants or strains of the same variety.

   (b) Inter-varietal or intra-specific:– Such crosses are made between two varieties of the same species: this is the most important category of crosses which have produced most varieties of crop plants such as maize, sorghum, rice, etc.

   (c) Inter-specific or intra-generic:– Such crosses are made between different species of the same genus. These are practised if the desired characters are not available in the same species. Usually these are needed for stress resistances such as diseases, pests, drought resistance, etc.

   (d) Inter-generic:– Inter-generic crosses are made between plants belonging to different genera, for example, crossing of wheat (*Triticum aestivum*) and rye (*Secale cereale*) to develop triticale. Inter-specific and inter-generic crosses are also known as wide crosses.

(ii) **On basis of Mating System:**

   (a) Single cross:– When two parents are crossed to produce hybrids, this is known as single cross, for example, A × B, A and B may also be crossed using B as female to give B × A cross. This is known as reciprocal cross. Single crosses are used either for making selection in F2, or for making crosses with other parents or for release as hybrid variety.

   (b) Three way cross:– As the name indicates, three way cross involve three different parents: A × B – F1 × C

   (c) Top cross:– This involves crossing lines or clones, etc, to a topcross parent (tester) which may be a variety, inbred line or single cross. Depending on the parents involved, this is also known as inbred variety cross, double top cross or a three way cross.

(d) Double cross:– Double cross involves crossing of two single crosses (A × B) and C × D as (A × B) × (C × D).

(e) Back cross:– Back cross is the repeated crossing of F1 and resulting hybrids with one of the parents:
A × B → F1 × A = BC1;
A × B → F1 × B = BC2.

(f) Polycross:– It is the intercrossing of a group of cultivars or clones in isolation so as to allow random pollination.

(g) Synthetic cross:– Synthetic cross results by random pollination of 4 – 10 pre-tested lines in isolation.

These types of crosses have been discussed in detail elsewhere.

(h) Multiple cross: recombination possible from single crosses are too restrictive to permit rapid improvement in self-pollinated crops. Hence, Harlan et al. (1940) proposed a mating system where a number of varieties or pure line (up to 32) are crossed in successive generations into single crosses, double crosses, octuple crosses (Figure 3.9) until the final hybrid involves all parents. Theoretically the multiple cross provides opportunity for recombination among genes from many parental strains. However, there are two limitations. First, it is not practically feasible to obtain enough F1 seeds to retain all parental genes in the final crossing generation. The second practical limitation is that it is not always possible to find 16 or 32 good parents and there is likelihood of poor parents being used. Further selection in the material may be made either by the pedigree or bulk method. Multiple crosses are made primarily to combine several characters from a number of parents at the same time, or for developing multiline varieties (see Chapter 6).

5.3 Hybridisation Technique

The hybridisation technique consists of the selection of parents, system of pollination control, pollination and collection of F_1 seeds. The selection of parents depends on the aims and objects of breeding. System of pollen control is exercised in breeding programmes. In nature, either in self or cross pollinated crops, pollination and fertilisation occurs in natural way. But in hybridisation programme, fertilisation has to be done by pollen from a desired plant or parent. Thus it becomes necessary for pollen control to remove the male flowers before pollination. There may be different systems for it depending on the type of plant and the morphology of the flower.

(i) **In monoecious plants:** These plants have unisexual, that is male and female flowers are separate but on the same plant; for example,

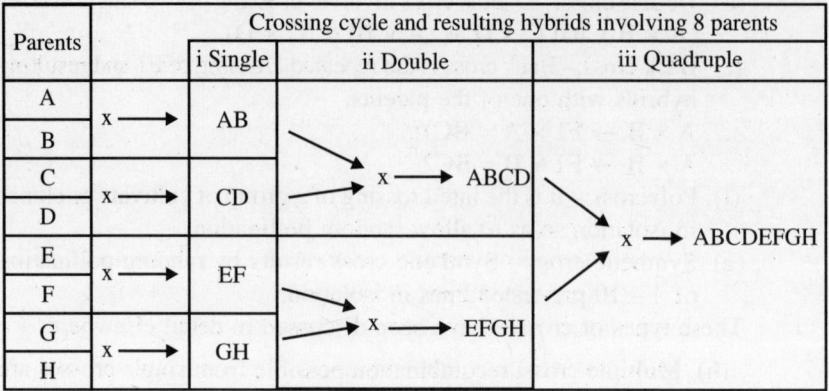

Parents		Crossing cycle and resulting hybrids involving 8 parents		
		i Single	ii Double	iii Quadruple
A	x⟶	AB		
B				
C	x⟶	CD	x⟶ ABCD	
D				x ⟶ ABCDEFGH
E	x⟶	EF		
F				
G	x⟶	GH	x⟶ EFGH	
H				

Figure 3.9   A system of Multiple Crossing involving 8 parents and the resultant hybrids.

maize, cucumber, pumpkin, coconut, etc. Emasculation in this group of plants is done by removing the male flowers. In maize, male inflorescence is called tassel and there for emasculation is done by removal of tassel, that is, detasseling.

(ii) **In dioecious plants:**   Male and female flowers are borne on different plants, for example, *parwal*, hops, hemp, date palm, spinach, asparagus, etc. Pollination may be controlled by removing the male plants from the population.

(iii) **In plants with bisexual flowers:**   In this group, all flowers have male and female parts and thus removal of the male plant or male flower is not possible and removal of the male part of each flower becomes essential. Removal of the male parts from a flower is known as emasculation. Three major methods may be applied to remove male parts or render it useless.

6. EMASCULATION

Emasculation is the removal of male part from a flower so that pollination could be done by using only desired parent. After removing the male part, flower (female) is covered with a bag to avoid any cross pollination by air or insects.

6.1 Physical Methods of Emasculation

(a) **Emasculation by hand:**   Male part of the bisexual flowers like rice, wheat, beans, tomato, brinjal and others, male part of the flower (anthers) are removed by using forceps and hand. Hand emasculation is usually done in the evening hours considering that the time of anthesis is in the morning hours. During night hours any injury to the emasculated flower may also heal. Emasculation is also done by machine i.e. use of Vacuum Emasculator (Figure 3.10) in rice and other small flowered plants. Anthers get sucked out of the flower under vacuum created by the machine. This does not injure the flowers.

(b) **Forced or cut opening method:**   This consists of opening the flower by forceps and removing the anthers as in the case of rice, chickpea, pea, jute, potato, tomato, etc. The part of floret may also be cut to expose anthers and these may then be removed as in case of wheat or rice. In linseed, the stamens are removed by pulling out the petals. In brinjal and tobacco, the stamens are removed with the fingers. A vacuum emasculator has also been developed to remove anthers (Figure 3.10). The exact technique depends on individual crop but summarily it is the exposure and removal of anthers with minimum injury to flowers.

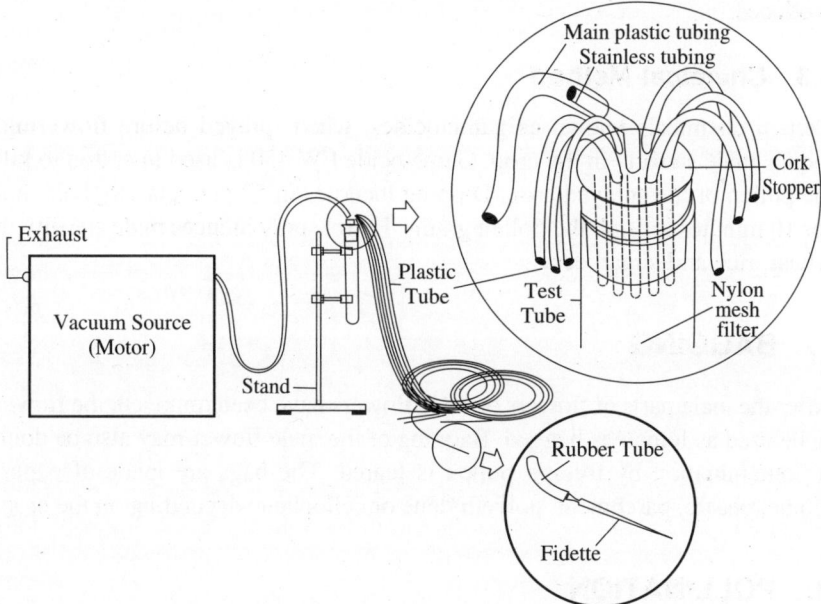

Figure 3.10 Emasculation using Vacuum Emasculator. Pipettes suck the anther out of flowers, as these are connected to a vacuum source. Up to 6 people can work simultaneously using one machine.

(c) **Induced opening method:**   Covering the inflorescence in black paper envelope or dipping in hot water (45 to 53 C degree) for 1 to 10 minutes as in the case of rice induces the premature opening of flowers. Once unburst anthers protrude out, these are removed by scissors. In sorghum and many grasses hot water treatment as above kills the pollen, without injuring the ovule.

(d) **Removal of male flower:**   In the monoecious plants like most cucurbits, maize etc, this is done by removing the male flowers from the plants leaving only female flowers. In maize this is done by removing the tassel and is known as Detasseling.

6.2   Genetic Method

In some self-pollinated crops such as barley, sorghum, onion, bajra, wheat and rice emasculation may be done by the use of male sterile lines. Male sterility once introduced eliminates the emasculation process. Similarly, self incompatibility can also be used. Male sterility has been discovered in many crop plants and is being used for large scale crossing and production of hybrid seed commercially. Rice is a good example where large scale hybrid seed is produced.

6.3   Chemical Method

Certain chemicals known as gametocides, when sprayed before flowering, induce male sterility in the crop. Gametocide FW 450 is used in cotton to kill the pollen of the sprayed crop. Dipping lucernes in 57 per cent ethyl alcohol for 10 minutes kills all the pollen grains. Ethrel spray induces male sterility in wheat, rice and sugarbeet.

7.   BAGGING

After the male parts of flowers or male flowers have been removed, the flower to be used as female is bagged. Bagging of the male flower may also be done if contamination by foreign pollen is feared. The bags are made of paper, clothe, plastic, parchment, polyethylene or cellophane depending on the crop.

8.   POLLINATION

Pollens or anthers are collected from the desired male parent and applied on the stigma of the female flower after opening the bag. Pollen either collected in bag or direct from dehiscing inflorescence may be dusted on the stigma as done in maize and wheat respectively. In some, instead of pollen, mature anthers are inserted in florets with the help of forceps.

In crops like bajra, jowar and sugarcane, hand pollination is difficult due to the small size of flowers. In such crops the male parent is either planted by the side of the female or the inflorescences are covered in one bag to effect pollination.

Pollination is done usually in the morning hours of the day following emasculation. Usually this coincides with the anthesis time when the stigma is receptive and pollens dehisce. After pollination, the female parent is re-bagged and tagged with the name of the female and male parents and the date of pollination.

(i) **Raising F_1 plants**

On maturity the seed, which develops after crossing, that is, the F_1 seed, is harvested, dried and stored. Certain seeds such as of sugarcane start losing viability and hence need to be sown immediately. In crosses involving up to two true breeding parents, the F_1 plants are exactly similar. In such cases any plant looking similar to the female parent or of different morphology may be a self and hence should be rejected. Different types of F_1 plants may be expected in multiple crosses and crosses involving heterozygous parents. Rejection of self plants in such cases is difficult and hence the seed of every dissimilar plant should be collected separately. The F_2 population of these plants is also, grown separately.

(ii) **Barriers to Crossability**
   (a) Geographical separation
   (b) Non flowering
   (c) Non synchronisation of flowering time
   (d) Incompatibility
   (e) Failure of embryo to develop
   (f) Sterility of hybrid
   (g) Wide crosses.

The parents to be crossed, if located, at different places, have to be brought together. But sometimes the introduced plant does not flower at the new place owing to photo-period, temperature and soil differences. By manipulation of the appropriate factor the plant may be made to flower.

If flowering time does not synchronise, crosses cannot be made, and storage of pollen using lower temperature and lower humidity is done. Treatment with hormones may also be made to hasten or delay the flowering. Staggered planting is also done to synchronize the long duration late flowering varieties with the ones from the first date with early or later date.

Incompatibility mechanisms become a barrier in many cases. These may be removed by choosing appropriate parents, manipulation of stigma and style, hormonal treatment, bud pollination or delayed pollination, etc.

After fertilisation has been affected, the embryo may fail to develop usually in inter-specific and inter-generic crosses. For example, in wheat and rye crosses, the embryo does not develop beyond 20 days. Therefore, in order to save the embryo, it may be removed from the developing seed and grown in an embryo culture much earlier. The excised embryos are cultured in artificial media till roots and leaves differentiate. There after the young seedlings are transferred to the soil.

Sterility of hybrids is a major problem in crosses involving distantly related parents or parents with unequal chromosome number. These require specialised techniques. One such technique is protoplast fusion.

(iii) **Protoplast fusion and cell hybridisation**

Wide crosses are not successful, but are now possible with the help of protoplast fusion. The first success in protoplast fusion was achieved by Harris and Watking (1965) when they were able to fuse somatic cells of man and mouse. Fusion of plant cells is not very easy because they are bound by rigid cellulosic wall and the adjacent cells are held together by protein rich substances. The discovery of enzymes that facilitate the dissolution of cellulosic and pectic substances helped in the isolation and fusion of plant protoplasts. Takebe (1971) and Nitsch (1972) and their colleagues were able to isolate protoplasts of tobacco and could regenerate entire tobacco plants from it. Carlson (1972) has been able to raise complete hybrids of *Nicotiana glauca* and *N. longsdorffi* through the technique of protoplast fusion. The application of this technique in incompatible and wide crosses gives a new tool in the hands of the plant breeder.

(iv) **Methods of Breeding and Mode of Reproduction**

Ever since the domestication of crops, man started growing the strains, which he had found superior with his experience. The modern plant breeder also does the same job, that is, provides varieties superior than the existing ones using a number of plant breeding methods. Basically, the breeding of any species of crop plants, regardless of its natural mode of reproduction, involves two phases:

(i) The creation of a reservoir of genotypic variation.

(ii) Selection among the genotypes.

A reservoir of genotypic variation can be created by collection of germplasm from various sources, hybridisation among varieties, species and even genera and mutation (Table 3.1).

Table 3.1 Sources of variation used in plant breeding

A.	Natural populations;	
	(i)	Cultivars, ecostrains, species, etc.
	(ii)	Introductions of other such germplasm
B.	Hybridisation:	
	(i)	Inter-specific or inter-varietal
	(ii)	inter-specific or inter-generic
C.	Induced mutation:	
	(i)	Major genic
	(ii)	Polygenic
D.	Induced recombinations:	
	(i)	Breakage of close linkage
	(ii)	Reconstitution of chromosomes
E.	Polyploid:	
	(i)	Haploid
	(ii)	Triploid
	(iii)	Tetraploid
	(iv)	Other ploids

Selection may then be practised for identifying superior types from the so created population. The methods of creating genotypic variation and the systems of selection are called methods of breeding. The methods depend on the nature of crop and mode of its pollination. Therefore, separate considerations are needed for individual crop category.

9. METHODS OF BREEDING SUMMARISED

The methods of breeding are listed below according to the mode of reproduction in crops.

9.1 Self Pollinated Crops

1. Introduction and acclimatisation
2. Selection

    (i) Mass selection
    (ii) Pureline selection
3. Hybridisation and selection using
    (i) Pedigree method
    (ii) Bulk method
    (iii) Single seed descent method
    (iv) Back cross method
4. Heterosis breeding
5. Mutation breeding
6. Ploidy breeding
7. Innovative methods of breeding

9.2 Cross Pollinated Crops

1. Introduction and acclimatisation
2. Selection
    (i) Mass selection
    (ii) Inbred line selection
    (iii) Simple recurrent selection
    (iv) Reciprocal recurrent selection
3. Hybridisation and polycross for sound selection
4. Heterosis breeding using
    (i) Single cross hybrids
    (ii) Double cross hybrids
    (iii) Top cross and three-way cross hybrids
    (iv) Synthetics and composites
5. Population improvement
6. Mutation breeding
7. Ploidy breeding
8. Innovative methods of breeding

9.3 Asexually Propagated Crops

1. Introduction and acclimatisation
2. Selection
    (i) Clonal selection
    (ii) Mass selection
3. Hybridisation and selection
4. Mutation breeding
5. Ploidy breeding
6. Innovative methods of breeding

These methods have been discussed in next 9 chapters in detail illustrated with examples. Comparative use of these methods in three groups of crops is summarised in Table 3.2.

Table 3.2 Comparison of use of breeding methods in self pollinated, cross pollinated and asexually propagated cops

Breeding method	Self pollinated	Cross pollinated	Asexually propagated
Introduction	Yes	Yes	Yes
Selection:			
Mass	Occasionally	Yes	Occasionally
Pureline	Yes	For inbred lines	No
Recurrent	Rarely	Yes	No
Clonal	No	No	Yes
Hybridisation and selection:			
Pedigree	Yes	No	Rarely
Bulk	Yes	Yes	No
Back cross	Yes	Yes	Rarely
Hybrid breeding	Occasionally	Exclusively	Yes
Ploidy	Rarely	Rarely	Yes
Mutation	Yes	Rarely	Yes
Innovative	Yes	Yes	Yes

10. USEFUL REFERENCES

Chopra, V. L. (Ed.) 2000. *Plant Breeding: Theory and Practice.* Oxford & IBH, New Delhi.

Cobley, L. S. 1963. *Introduction to the Botany of Tropical Crops.* Longmans, Green and company Limited, London.

Crane, M. B. and Lawrence, W. J. C. 1952. *Genetics of Garden Plants.* The McMillan & Co., London.

Elliot, F. C. 1958. *Plant Breeding and Cytogenetics.* McGraw-Hill Book Co. Inc. New York 395.

Fehr, W. R. And Hadley, H. H. 1980. Hybridisation of Crop Plants. Amer. Soc, Agron. Madison.

Fryxell, P. A. 1957. Mode of reproduction in higher plants. Bot. Rev. 23: 135.

Hutchinson, J. B. 1958. *Genetics and Improvement of Tropical Crops*. Cambridge University Press, Madison.

Khokholov, S. S. (Ed.) 1976. *Apomixis and Breeding*. Amerind Publ. New Delhi, 346.

Maheshwari, P. 1950. *An Introduction to the Embryology of Angiosperms*. McGraw-Hill Book Co. Inc. New York 453.

Sneep, J. And Hendriksen, A. J. T. 1979. *Plant Breeding Perspectives*. Pudoc, Wageningen, Netherlands.

Wallace, B. 1963. Mode of reproduction and their genetic consequences. In: Statistical Genetics and Plant Breeding. National Academy of Science, Washington, D. C. 982.

Williams, W. 1964. *Genetical Principles of Plant Breeding*. Blackwell Sci. Publ. Oxford.

Origin, Domestication and Introduction of Crop Plants

Various species of plants arose through evolution from less evolved to more evolved forms. The progenitors of the present day crop plants were wild. How long it has taken the wild species to become like cultivated ones is a matter of guess. The evidences from archaeology are not older than 6,000 years. Anyway, it is assumed that with the genetic variability provided by gene mutation, recombination, inter-specific hybridisation and polyploidy, natural selection continues to evolve undomesticated forms. Prehistoric agriculturists domesticated these forms and provided human selection pressure to evolve the cultivated forms.

It is assumed that cultivated plants originated in a particular area only. The determination of the place of origin of crop plants has been done by collection of existing forms of cultivated and wild species. The Russian scientist N.I. Vavilov and his colleagues collected a wealth of material from across the world and studied them over a period of over 10 years. Vavilov found that the entire variability of crop plants collected by his team is centred in eight regions of the world. Vavilov (1951) named these eight regions of the world as the **Centre of Origin of the Crop Plants** (Figure 4.1).

1.   CENTRES OF ORIGIN

1.1   Chinese Centre

The area consisting of Central and Western China(1) is considered to be the most important centre. It is primary centre of origin of soybean, radish, *Colocasia sp.*, buckwheat, poppy, apricot, peach, litchi, orange, Chinese tea (*Camelia sinensis*), etc, and are believed to have originated here. It is a secondary centre for waxy maize, cowpea, turnip, sesame and turnips also.

1.2   Hindustan Centre

Part of India excluding Punjab but including North-East India, Myanmar, Malaysia, and major Islands of Indonesia(2) are said to the centre of origin of rice, sugarcane, pigeon pea, chickpea, mung bean, cowpea, lobia, brinjal, hemp, black pepper, radish, cotton (*Gossypium arboreum*, and *G. herbaceum*) turmeric, indigo, banana, sour lime, mango etc. A sub centre 2a, Indonesian centre, was also identified where sugarcane, banana and coconut is said to have originated.

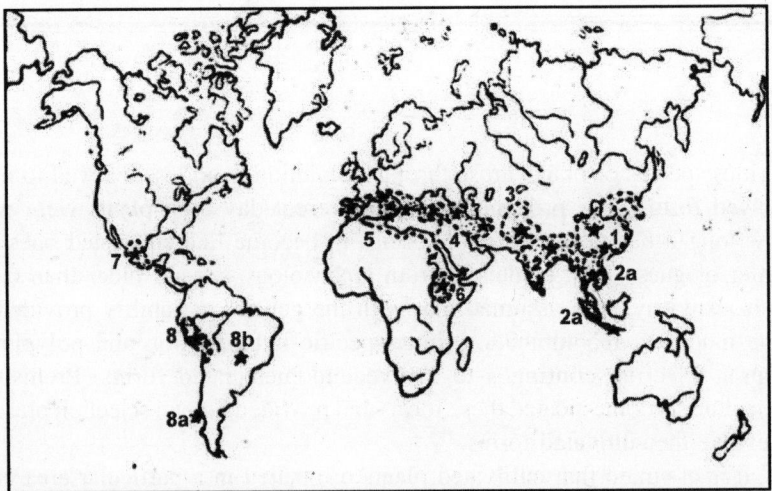

Figure 4.1   Centres of Origin of Crop Plants. For the explanation of numbers, see text.

1.3   Central Asiatic Centre

This includes North-West India (Punjab, Kashmir), Pakistan, Afghanistan and South-Western USSR(3). Crop plants like almond, apricot, apple, broad bean (*Vicia faba*), bread wheat (*Triticum aestivum*), club wheat (*Triticum compactum*), carrot, cotton (*Gossypium herbaceum*), garlic, grape, lentil, linseed, musk melon, onion, pea, pistachio nut, pear, sesame, safflower, spinach are believed to have originated here. It is a secondary centre for rye (*Secale cereale*) also.

1.4   Near-Eastern Centre

Known also as Asia Minor Centre or Persian Centre, this centre includes countries of the middle-east like Turkey, Iran, Israel, and whole of Transcaucasia(4) region. Plants like alfalfa, clover (*Trifolium sp.*), two row barley, 9 species wheat, linseed, oat, some species of *Allium*, apple, almond, fig, grape, pomegranate etc, originated here. It is secondary centre of origin of rape (*Brassica campestris*), black mustard (*Brassica nigra*), and turnip.

1.5   Mediterranean Centre

Located in countries around the Mediterranean sea(5), crop plants like barley, beans, durum wheat (*Triticum durum*), emmer wheat (*Triticum dicoccum*), barley, broad bean, cauliflower, cabbage, several species of oat, several species of *Lathyrus*, lentil and sugarbeet originated in this centre.

1.6   Abyssinian Centre

Embracing Ethiopia and Eritrea(6), this centre is primary centre for barley, several species of wheat (*T. durum, T. dicoccum, T. turgidum*), Bajra, Sorghum, safflower, coffee, lady's finger, sesame etc. It is secondary centre of origin of broad bean (*Vicia faba*).

1.7   Central American Centre

Rather an isolated part in south Mexico and neighbouring countries in Central America(7), this was considered as the centre of origin of important crops like maize, beans (*Phaseolus vulgaris*), chillies, cotton (*G. hirsutum*), pumpkin, papaya, guava and gourd.

1.8   South American Centre

This includes Peru, Ecuador, Bolivia, Chile, Brazil, Paraguay and neighbouring islands(8). Crops like Egyptian cotton, tomato, tobacco, sweet potato, papaya, cassava and cashew nut are believed to have originated in this centre. Out of

South American Centre, Vavilov separated two sub centres, **8a, Chiloe Centre** where potato originated; and **8b, Brazilian-Paraguayan Centre** where groundnut, *Hevea* rubber, pineapple and cashew nut originated.

2.   CENTRE OF DIVERSITY AND CENTRES OF ORIGIN

All cultivated plants have descended from wild ancestors in the remote past. The phenomenon has taken place in particular geographic areas. Vavilov (1951) identified these areas by studying the genetic diversity in a particular crop species. The areas where he found maximum diversity of forms he called the centre of origin for that plant species. He recognised two types of centres: Primary centre of origin, a place of diversity where the varieties grown have maximum number of dominant genes; Secondary centre of origin, where two or more species crossed together and where natural and artificial selection occurred subsequently. For example, on above considerations he identified Mexican as primary centre and Chinese as secondary centre of maize. With increasing knowledge of crop geography and diversity, in some sporadic cases the centre of diversity is not coinciding with the centre of origin. For example, Ethiopia has tremendous diversity in barley but is not its centre of origin. May be barley was introduced in Ethiopia later but due to agroclimatic diversity or pressure of domestication, genetic variability got created and preserved.

Harlan (1951) called such small areas 'gene micro-centres', where evolution is still proceeding at rapid rate as in Turkey where he recognised gene micro-centres for wheat. Zhukovasky (1968) enlarged Vavilov's centres to include the area of domestication also and called it megagene-centre. He added four new centres: Australia, Africa, Siberia and North America to the existing eight and thus identified 12 megagene centres where crop plants originated and were domesticated (Zeven and Zhukovasky, 1975).

2.1   Domestication

Domestication is the taming and training of plants to suit the human needs and putting them under cultivation. In other words, domestication of plants is the change of the ideotype to adopt them better to man-made environments. Plant domestication, human civilisation and agriculture are closely related phenomenon. *Thus domestication is the process of bringing wild species under human management.* The early civilisations of the Incas in South America and Mayas in Mexico developed around maize, and of the Babylonians and Egyptians around wheat and barley. Inhabitants of these places were obviously involved in domesticating these crops. Thus domestication also must have passed through the periods of proto agriculture and incipient agriculture to effective agriculture.

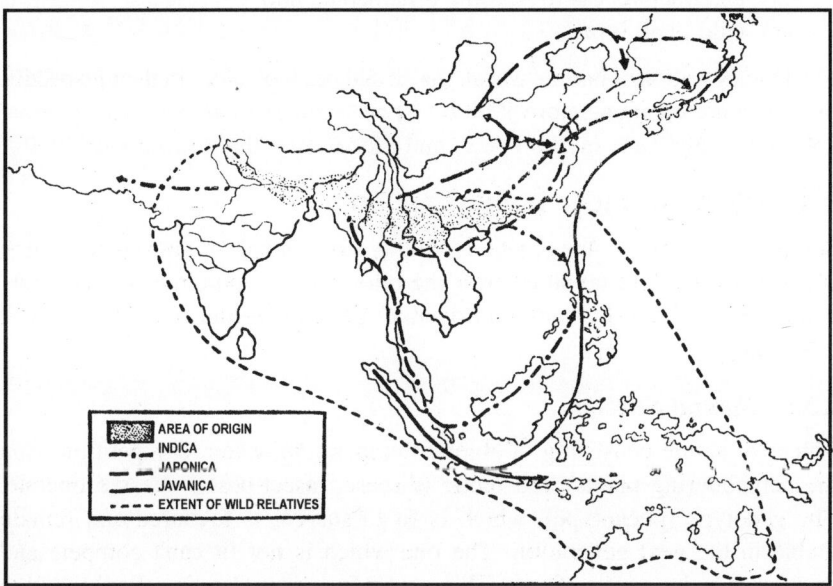

AREA OF ORIGIN
INDICA
JAPONICA
JAVANICA
EXTENT OF WILD RELATIVES

Figure 4.2 Origin of rice and dispersal of its races (Source: IRRI Reporter 1/78)

2.2 Effects of Domestication

All cultivated plants today were wild some day. The wild forms had many undesirable characters referred as wild characters against which human selection proceeded. Thus more evolved forms arose for various agroclimates and purposes. Due to cultivation some plants have changed very quickly and these are used in diverse ways. For example, Brassica is now cultivated not only for oil, but for vegetable, fodder, ornamental, walking sticks and construction material. Maize, a starch crop, is now exploited as an oil crop. Clearly then, mutation, hybridisation and gene recombination under the influence of human selection have led the cultivated plants to change or lose one or more of the following characters to become better cultivated types:

   (i) strict environmental requirement,
  (ii) synchrony in flowering,
 (iii) shattering and seed dispersal mechanism,
 (iv) seed and fruit size,
  (v) seed dormancy,
 (vi) photo periodic controls,

(vii) pollination habit,
(viii) defensive adaptation as hairs, thorns, etc,
(ix) unattractiveness to become ornamental, and
(x) wild plant type.

Domestication of barley, wheat, maize and pea took place in their respective place of domestication before 5000 B.C; potato, rice, sorghum, cotton in before 2500 B.C; sugarcane in 1000 A.D; and rubber after that (Simmonds, 1976).

2.3 Selection Under Domestication

Selection operates over the genetic variability. For example in a mixed population some types may get favoured over the other but in a pureline, no selection, natural or artificial (human), can operate. Let us consider both the types of selection.

2.3.1 *Natural Selection*

Forces of nature consisting of abiotic (temperature, water, competition from the neighbouring plants) and biotic (disease, insect-pests) stresses operate. The genotype (phenotype), which is fit to survive is favoured and remain viable in the next generation. The one which is not fit can't compete and disappears in due course of time. This is what in a sense Darwin called evolution or survival of the fittest.

2.3.2 *Artificial (human selection)*

Artificial selection on the natural population involves the choice of mankind. What they desire from a particular plant population. Depending on the choice of the selector it could be:

(a) **Directional selection** leads to a population one extreme direction. It could change the phenotype or mode of reproduction of the population in the direction of the selection. If it is for large seed size, the resulting population could have large seed size. The other example is from tomato which in its place of origin (Peru - Bolivia) was cross pollinated but under European glass house domestication it became self pollinated.

(b) **Stabilizing selection** leads not to extreme type but to intermediate phenotypes. The stabilizing selection favours those genotypes whose phenotypic expression clusters around the population mean. Extreme types are not favoured.

(c) **Disruptive selection** operates on those populations that are subjected to distinct ecological niches. In each ecology, a different phenotype gets selected so that the population ultimately consists of more than one recognizable form. Thus disruptive selection maintains polymorphism in a population.

2.4 Domestication of Some Crops

(i) **Rice (*Oryze sativa* L.)**

Most probably rice originated in the 3rd millennium B.C. in South and South East Asia. The rough area may be marked, in the north bordering Himalaya, in the south up to Deccan plateau and in the east touching parts of Myanmar, Thailand and Vietnam (Figure 4.2). The progenitors of rice originated in the tropical swamp and were shade-loving. Cultivated forms arose out of some wild species resembling wild rice *O. perennis*. There are others who believe that rice ($2n = 24$) is balanced secondary polyploid with a basic chromosome number of 5. This number got duplicated and two additional chromosomes added ($10 + 2$). This plant on duplication gave rise to the progenitor of rice with $2n = 24$. From India rice moved to China where probably for the first time it was domesticated. It moved northward to Japan via Korea. From India, rice moved to Africa and America through the Arabian countries. During this process of domestication *O. sativa* differentiated into three forms: *indica, japonica,* and *javanica.* Japonica is cultivated in North china, Korea and Japan, and Javanica in Indonesia. In the rest of the world indica is cultivated. The only other cultivated species of rice *O. glaberrima* is limited to West Africa.

During domestication, many morphological and physiological changes have gone in, starting from adaptation to habitats of normal lands and open sun from the shady swamps. Incorporation of non-shattering character and absence of dormancy has gone in. Differentiation of *O. sativa* into three forms has been a result of change in leaf shape and size, grain shape and hairiness and other plant structure. Various types even in *indica* differentiated which inhabited upland, medium land and deep water conditions. In the past 20 years plant breeders have changed the entire plant structure based on the 'plant ideotype concept'.

(ii) **Wheat (*Triticum aestivum* L.)**

Having originated in the mountains of Afghanistan and South-Western Himalaya, it moved with man westwards and eastwards. Wheat has been in cultivation since over 6000 years in this area. Four species of wheat are cultivated in India: *T. aestivum* is cultivated in the maximum area and is used for *chapati* making. *T. durum* is cultivated in limited dry areas of Maharashtra, Madhya Pradesh and Gujarat. *T. dicoccum* is grown in very limited black rust affected areas of Mysore, Telangana and Maharashtra. The areas of cultivation of *T. turgidum* are negligible. The major change that has gone under domestication was the non-brittle rachis. Of late the plant type has been changed by breeders into one which is more management responsive. Thus a wheat plant of the 1970s looks entirely different in leaf canopy and plant habit than of the 1950s.

(iii) **Barley (*Hordeum vulgare* L.)**
Barley originated in an area between the Mediterranean Sea and Afghanistan. Its wild progenitor is wild barley (*H. spontaneum*), which is two-rowed and has brittle rachis. Its domestication took place in 7000 B.C. and gave way to two rowed barley. The change from the brittle rachis of the wild to the tough rachis of the cultivated variety would have taken place between 7000 B.C. and 6000 B.C. Six row barley and other cultivated forms arose during domestication.

(iv) **Maize (*Zea mays* L.)**
Maize originated in South America (Mexico). The parentage of maize has been much disputed but the present conclusion is that it originated from teosinte (*Zea mexicana*). It is believed that teosinte (*Zea mexicana*) was cultivated in 5000 B.C. and tripsacoid maize in around 4000 B.C. and maize in 3600 B.C.

From Central America maize reached South America where it became agriculturally more advanced. The oldest domesticated varieties had slender cob but later one got thicker cob. Other morphological changes due to domestication are complete husk coverage of cob, reduction of the glumes of inflorescence, arrangement of spikelets in higher row number and increase in the silk length. Introgression of teosinte genes in maize had been there and is still going on.

Portuguese traders took maize to India in the 16th century. A secondary centre arose in Sikkim and bordering areas in China, which Vavilov recognised as the secondary centre for waxy maize. Several types of maize evolved and are in cultivation now like pop, sweet, soft, baby, dent, flint, etc, which reflects the diverse combination of kernel size texture and starch sugar ratio.

(v) **Pigeon pea (*Cajanus cajan* Mille)**
Its African origin was described by de Candolle (1886). But it is typically of Indian origin as reported by Vavilov (1951) and supported by many. Pigeon pea is believed to have originated from its wild progenitor *Cajanus cajanifolius*. Pigeon pea was domesticated in India

and from there it moved to Africa and America. Eastward it moved to Malaya, Indochina and Australia.

(vi) **Chickpea (*Cicer arietinum* L)**

Chickpea moved from its ecological optimum, the foothills with high light intensity, long days, and well-drained soils of region around Turkey to the plains of India where it changed itself and adapted well under domestication. De Candolle (1886) believed the chickpea to have originated between Greece and the Himalayas. Vavilov (1951) postulated two centres of diversity, one primary centre in India, which is now recognised as the centre of origin. The other one in Ethiopia is a secondary centre. In Turkey probably it was under cultivation around 5450 B.C. (Helback, 1970). In India, it is under cultivation since 2000 B.C. White Kabuli gram arrived in India in A. D. 1700.

(vii) **Pea (*Pisum sativum* L.)**

It has central Asiatic origin. The domestication is as old as 6250 B.C. In Switzerland peas were in cultivation even in the Stone Age. Cultivated forms originated from wild species *P. elatius*. It later differentiated into garden and field peas. Its domestication may have taken place in Southwest Asia and spread to India through the Himalayas and Tibet (Smart, 1976).

(viii) **Lentil (*Lens esculanta* Moench)**

This has also Central Asiatic origin. The cultivated species probably originated from wild species *L. nigricans* and *L. orientalis.* The area of first domestication was in the Near East. Agro-ecological groups meet in turkey where micro centres have developed.

(ix) **Cotton (Gossypium sp.)**

Chiefly four species of cotton are under cultivation, two of them diploids (and two others tetraploid. Out of these *Gossypium arboreum* L. and G. *herbaceum* L. which are diploids originated in India and were domesticated here. The Indus Valley civilisation which collapsed in 2000 B.C. had played an important role in its domestication. The Hindus were first to weave cloth out of its fibre. From India these two species moved to Africa via Arabian countries. G. *barbadense* L. (tetraploid) originated in tropical South America and from there it moved to North America and Africa. G. hirsutum L. (tetraploid) originated in Guatemala in south Mexico. It was domesticated there and then moved to North American and other countries. The major change that took place under domestication was increase in length of corboluted lint and its spinnability.

(x) **Sugarcane (*Saccharum officinarum* L.)**
Originated in India and was domesticated here. During domestication, the sugar content, rind thickness and juice content increased. By 327 B.C. sugar cane was an important crop in India. From here it spread to Egypt in A. D. 641. Spain in A. D. 755, West Indies in the 15th century A. D. and U.S.A. in A. D. 1741. Modern varieties of sugar cane are complex hybrids of three cultivated species, namely, *S. officinarum* L., *S. sinense* Rox B., and *S. barberi* Jesweit, and many wild species of Saccharum and related genera. The thick stemmed *S. officinarum* and colourful cane was in cultivation in tropical countries. India is considered as the centre of its domestication as it was cultivated for chewing. The thin stemmed *S. sinense* Rox B., and *S. barberi* Jesweit were also cultivated but for jaggery (*gur*) making.

(xi) **Sugarbeet (*Beta vulgaris* L.)**
Originated in the south coastal region of Europe, and the Mediterranean region, it was derived from the wild species *B. maritima*. Further introgression of *B. maritima* and hybridisation with other wild species gave rise to red table beet, Swiss chard and fodder beet. Sugarbeet probably developed in Poland out of cross of table beet and fodder beet. The variety 'Weisser Schlesicher Zuckerub' is the parent of all sugarbeet varieties. The major change under domestication has been for biennial growth habit, large root size and high sugar content. It was used for extracting sugar in A. D. 1747. In India it was introduced in 1915, and domestication was mainly for types adapted to tropical conditions.

(xii) **Tobacco (*Nicotiana* sp.)**
Two species are under cultivation. *Nicotiana tabacum* L. originated in South America (Argentina) from where it moved to other parts of the world. It arose as an amphidiploid between two wild species *N. sylvestris* and *N. tomentosa*. This was followed by a long period of evolution through natural selection and domestication to bring numerous cultivars. Another species, *N. rustica* L. originated in Peru. This may have happened in Argentina where both the species occur wild. In India, tobacco was introduced in 1508 by the Portuguese traders.

(xiii) **Tomato (*Lycopersicon esculentum* Mill.)**
Originated in the Peruvian Andes of South America, tomato was limited to the Peru-Ecuador area for long. It was first domesticated in Mexico. The Spaniards took it to southern Europe in the 16th

century from where it spread to the rest of Europe. While being domesticated under glass house condition, it changed its nature from cross pollination to self pollination. The most putative ancestor of cultivated tomato is var. cerasiforme. From Peru, this variety spread as a weed to Mexico where it got cultivated due to similarity with raspberry (*Physalis*). The primitive cultivated tomato crosses freely with another wild species *L. pimpinellifolium* from where it got genes for high sugar. Fruit size in tomato has increased many times during domestication. In India, tomato was introduced in the 19th century.

(xiv) **Cauliflower (*Brassica oleracea* var. *botrytis* L.)**
Originated from its wild ancestor colowert, a stout weedy perennial, cauliflower has a history of 2,500 years. It was reported to have originated in the island of Cyprus from where it moved to Syria, Turkey and Europe. Dodens made the first illustration of crude cauliflower in 1554. It was domesticated particularly in Italy and then got popularity in the rest of Europe by the beginning of the 18th century. Cauliflower was introduced in North India around 1822. For about a century (1822 to 1929) it was domesticated, and selection continued for early maturity and adaptation to hot humid condition. Due to continued selection, Indian cauliflower drifted towards annual habit without any change in its original self-incompatibility (Hill, 1976).

(xv) **Potato (*Solanum tuberosum* L)**
Originated in Chiloe centre and was domesticated there. The Spanish people introduced it into Europe in 1500. The Portuguese brought it to India around 1715. Under domestication many characters of potato changed, particularly duration, photoperiodic response, tolerance to higher temperature, plant and tuber characters.

(xvi) **Orange (*Citrus sinensis* L)**
Orange is native of South-eastern Asia. It was domesticated between 1500 to 1000 B.C. in China and reached India afterwards. In the 15th century it was carried from India to Europe by traders. The Spanish took it to Florida in 1565. In Northeast India tremendous variability has been reported. A sporadic centre exists in Brazil from where the famous variety 'Washington Navel Orange' was collected. Tremendous variability exists in citrus in Assam area also.

(xvii) **Tea (*Camellia* sp.)**
Tea is believed to have originated in the primary centre somewhere in Central Asia about 60 degree N. This further differentiated into

two types: small leafed China type and broad leafed Assam type. The China type (*Camellia sinensis*) further differentiated and was domesticated in South China. It was introduced into India in the early part of the 19th century. The Assam type (*C. sinensis assamica*) had a secondary centre in North-east India along the Brahmaputra river. It was under cultivation by tribes but its commercial use started in 1823 only.

3. PLANT INTRODUCTION

Man has carried plants with him wherever he has migrated and this has been one of the most important features in the development of agriculture vis-à-vis cultivated plants. *Plant Introduction is the process of introducing plants from their growing locality to a new locality.* Thus introduction is a dynamic process involving the change of place of plant from one to another, perhaps a state or country. Introduction of plants from a foreign country is called foreign introduction or exotic collection. Introduction from one state to another is called indigenous collection or indigenous introduction. With the introduction, is associated the process of acclimatisation. Acclimatisation is the adaptation or adjustment of an individual plant or population of plants under the changed climate for a number of generations.

3.1 Purpose

The following are the four purposes of introduction.

(i) **Use in agriculture, forestry and industry:**
New crops, plants or crop varieties are introduced from various places for use as food, fibre, wood, or medicinal or industrial purpose. Introduced material is also utilised by plant breeders for hybridisation work.

(ii) **For studying origin, distribution, etc:**
The distribution of plants and their various forms throw light on origin and evolution of crop or group of crop plants. Based on collections from different parts of the world, Vavilov gave the idea of centres of origin.

(iii) **For aesthetic interest:**
The group of ornamental plants are introduced for beautifying garden, parks, houses, etc. Most of the flowers and garden plants are introductions from far and near.

(iv) **For germplasm conservation:**
Germplasm is the collection of lines, clones, mutants, species, cultivars, land races, etc, of plants from as many sources as possible. With the spread of high yielding varieties the old varieties are going out of cultivation and thus are in danger of being lost. Similarly, due to construction of dams, clearing of land etc, the germplasm in those areas is in danger of extinction. Such germplasm is collected, introduced and preserved in living conditions as seed, plant or tissue.

3.2 History of Plant Introduction in India

The earliest recorded account of an organised expedition in the world is that of Queen Hatshepsut of Egypt who sent ships to East Africa in 1500 B.C. to procure incense tree. But in India there is no early record of introductions or expeditions except some scattered attempts. Planned attempts started since last 50 years (Singh and Hardas, 1975).

3.2.1 *Scattered Attempts*

Many types of plants were introduced by the early settlers, travellers, traders, missionaries and others introduced many types of plants. The Portuguese, who settled in Goa in 1510, are reported to be responsible for introduction of several crops such as maize, potato, chillies, groundnut, sweet potato, guava, papaya, pineapple, cashew nut, etc, in India. Similarly the British introduced some fruit crops and vegetables like tomato and cauliflower in latter part of the 19th century. In the beginning of the 20th century, when State Departments of Agriculture were established, introduction of newer varieties started vigorously. In the absence of any control or plant quarantine regulations many diseases and pests also got introduced along with the material.

3.2.2 *Planned Attempts*

Planned attempts started with the establishment of the Division of Plant Introduction at IARI, New Delhi, in 1946. Both, introduction from foreign countries and introduction from different states within the country were taken up systematically. The Nepal expedition in 1961, the Sikkim expedition in 1962 and the Assam expeditions of 1967-71 are the notable ones when a large number germplasm of various crops were collected.

4. INSTITUTIONAL ORGANISATION

Plant introduction and acclimatisation work is being carried out by the following organisation in the country.

4.1 National Bureau of Plant Genetic Resources (NBPGR)

Dr. B. P. Pal of the then Imperial (now Indian) Agricultural Research Institute (IARI) had approached the then Imperial (now Indian) Council of Agricultural Research (ICAR), in the 1930's to set up a unit for the assembly of global germplasm. The ICAR Scheme started functioning in 1946 in the Botany section of IARI with the late Dr. Harbhajan Singh as first operational scientist. The scheme was expanded into Plant Introduction and Exploration Organisation in 1956 and as a separate division of Plant Introduction in 1961. In 1976, it was further elevated to the status of the independent institute of ICAR, designated the National Bureau of Plant Introduction. In 1977 it was renamed as the National Bureau of Plant Genetic Resources (NBPGR) with its headquarter at New Delhi.

The NBPGR is now the central body for collection, introduction, organising expeditions, exchange and distribution of seed material and other plant propagules of agri-horticultural crops. NBPGR also added a long term storage facility in its National Gene Bank. By the year 2008, NBPGR has conserved 3,67,419 accessions in its gene bank. Out of these collections rice 86,210, wheat sorghum 19,430, pearl millet 7,989, and minor millet 20,458. Among major groups of collections NBPGR has 1,43,830 cereals, 52,261 millets and forages, 56,099 legumes, 53,319 oilseeds, 10,502 fibre and 23,685 vegetables.

The collected material is thoroughly described for plant characters, data is stored and handled with the help of computers. Since documented information on germplasm would be available with the computers, germplasm with any type of plant character can be quickly searched in the gene bank. Material would be stored in gene bank established at NBPGR for long term storage and future use. Besides, NBPGR also assesses the utilisation of introduced material, co-ordinates the work of other agencies and imparts training in plant collection, introduction and maintenance. NBPGR has 10 Regional Stations located at Akola, Bhowali, Cuttack, Hyderabad, Jodhpur, Ranchi, Shillong, Shimla, Srinagar, and Thrissur.

  (i) **Akola:** This substation carries out plant explorations in the central zone of India besides acclimatising and multiplying of introduced material for that zone.

 (ii) **Bhowali:** Located in the foothills of India, it has major responsibility to collect flora from that region.

(iii) **Cuttack:** Located in the central eastern part of India, it collects and catalogues major food crops of the region.

(iv) **Hyderabad:** Main centre for collecting, cataloguing crop germplasm from central and southern part of India. Major emphasis is on germplasm from semi-arid tropics.

(vi) **Jodhpur:** Exclusively meant for exploring and acclimatising plant material for the arid zone, this centre is located at the Central Arid Zone Research Institute.

(vii) **Ranchi:** This station is upgraded to a regional base station for the plateau area.

(viii) **Shillong:** This is recent addition to the regional centres. This centre has been created for the collection of germplasm from North-east India which has been reported to be a reservoir of genetic variability of many plants including rice, citrus, maize, etc.

(ix) **Shimla:** This station carries out explorations for germplasm collection in the northern hills. Acclimatisation of material introduced from temperate countries and high altitudes is also done here. It has major responsibility of temperate fruits and hill crops.

(x) **Srinagar:** Responsible for leading expedition and collecting and maintaining crop germplasm from the higher hills and valleys.

(xi) **Thrissur:** Leads and collects crop germplasm from southern part of India including islands.

4.2 Forest Research Institute, Dehradun

The plant introduction organisation set up at the institute looks after the introduction, acclimatisation and testing of forest trees. It also looks into the conservation of various forest trees.

4.3 Botanical Survey of India

Established in 1890, this body continues to introduce medicinal plants and plants of botanical importance.

4.4 Others

Some central research institutes also collect, introduce and maintain plants for their use; for example, for rice (Central Rice Research Institute, Cuttack), Sugarcane (Sugarcane Breeding Institute, Coimbatore), tobacco (Central Tobacco Research Institute, Rajamundry), potato (Central potato Research Institute, Shimla), Coconut (Central Plantation Crop Research Institute, Kasaragod), and other crop oriented research directorates (see Chapter 14), and pigeon pea, chick pea, groundnut (ICRISAT Hyderabad).

5. PROCEDURE OF INTRODUCTION

All plant introductions are co-ordinated by NBPGR. The user of the material, if s/he knows from where to introduce the material, may write direct to the scientist or institutions to send the material to him/her through NBPGR after obtaining an Import Permit from NBPGR. In case s/he does not know from where to introduce the material, s/he may request the Director of NBPGR to procure it. In either case, on receipt of the material it will be examined for the presence of pathogens and pests as per the rules and regulations set for Plant Quarantine in India. If the material is found free of these, it is given to user. A similar process has to be undergone when a material is to be exported to other countries. Along with a Phytosanitary (plant health) Certificate it is supplied to the user. These operations are done under plant quarantine regulations to ban the entry of any plant material carrying harmful organisms. The flow chart of introduction to utilisation and maintenance of germplasm is given in Figure 4.3.

Introduction to Plant Breeding

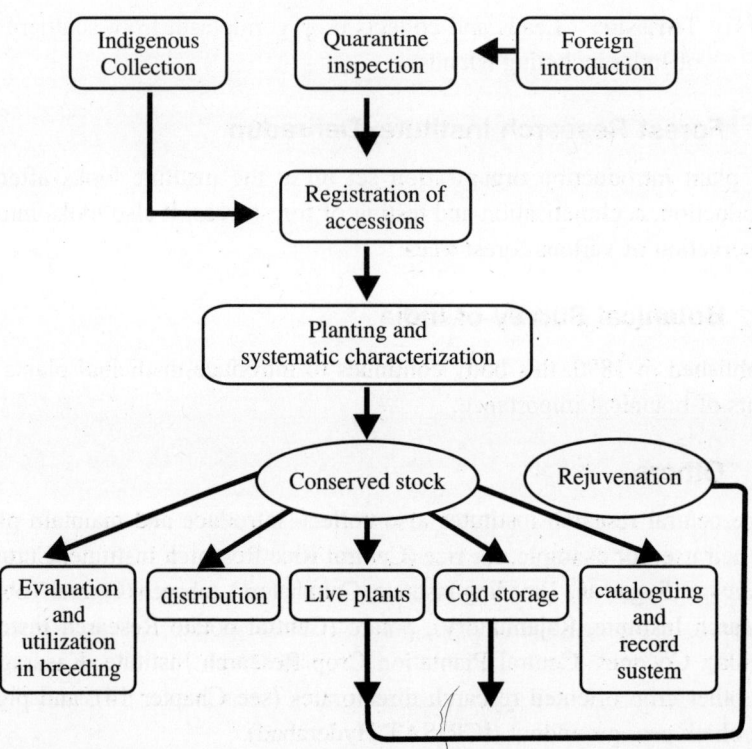

Figure 4.3   Flow chart of germplasm for utilization and maintenance.

Any plant propagule like seed, cutting, tubers, plants, seedlings, pollen grains can be introduced depending on the nature of the crop and the purpose of the user (Frankel and Bennet, 1970). There are several ways of introducing to a country.

5.1 Exploration

The probable place of collections and centres of diversity may be within or outside the country. Methods of exploration and germplasm collection have been given in detail by Bennet (1970). With the consent of the concerned countries, expeditions may be organised to collect material from natural habitats and introduce them to other places. In an organised expedition in North-east India, 5,528 rice germplasm was collected in 1967-71.

5.2 Exchange

The material may be obtained on an exchange (give and take) basis from friendly countries either direct or through USAID, Ford Foundation, international Institutions and Organisations, including FAO, and introductions from them may be made on an exchange basis. Some famous institutes which may be approached for seeds are: NI Vavilov Institute of Plant Industry, Leningrad (USSR); Plant Genetics and Germplasm Institute, Beltsville, Maryland, USA; International Crop Research Institute for Semi Arid Tropics (ICRISAT), Hyderabad; International Rice Research Institute (IRRI), Los Baños, Philippines; Division of Plant Industry, CSIRO, Canberra, Australia; CIMMYT (Internal Wheat and Maize Centre), Mexico; FAO, Rome, Italy.

5.3 Purchase or Gift

The material can also be purchased or obtained as free gift from individuals or institutions.

6. PACKING AND DESPATCH

Depending on the type of material, packing is done to insure safe arrival of the consignment. Seeds are cleaned and treated with fungicides to make them free of pests, weed seeds, other contaminants and seed borne diseases. Then convenient packing is made. Problems arise with non-dormant seeds, pollen, cutting and plants. Various methods are adopted to pack and dispatch the material so as to reach the destination in a viable condition (Whyte, 1956).

6.1   Entry and Plant Quarantine

The material may either be shipped to prescribed ports of entry, namely Bombay, Calcutta, Cochin, Madras, Nagapatanam, etc, or by air to Delhi international airport. On receipt, the material is put to detailed examination for the presence of insect, diseases, nematodes, weed seeds, etc. This is called, entry inspection, and is done by NBPGR and other recognised laboratories. If found unsuitable for import according to plant Protection and Quarantine Laws, these are either returned or destroyed. If found suitable, these are treated with insecticides, fungicides, nematicides and are released to users. The other method is post entry inspection, and is done by the inspection of crop raised from the introduced seed. This is done in isolation for all crops like groundnut and potato, which are known to carry viruses. The general objective of all 'quarantine and regulatory' measures is to prevent pests and diseases from entering this country, or, having entered, from spreading further. These measures are essential to stop some plant species like Mexican jumping beans or varieties from entering in the country.

6.2   Cataloguing and Release

On getting clearance from quarantine, the material is catalogued and then handed over to user.

7.   MAINTENANCE OF INTRODUCTIONS

The purpose of introduction and germplasm collections does not cease with the release of some varieties out of it, but the entire collection need to be maintained in live conditions. A central long term (for over 20 years) storage facility is established at NBPGR. But at it is also the responsibility of the user of the material to maintain it. The central crop research institutes as described in 'others' also maintain the collections of their interest.

The International Board of Plant Genetic Resources (IBPGR), with its headquarters at FAO, Rome, was established in 1974 to co-ordinated and formulate introduction and germplasm conservation on an international scale. Plant introductions and germplasm resources can be preserved in the following ways (Frankel and Hawkes, 1974).

7.1   Seasonal Growing

Germplasm collections of annual species may be grown every year and thus maintained in living conditions. But it is very difficult to maintain large collections as the method is prone to losses through non-germination, crop loss due to pests or abiotic stress, mixtures etc.

7.2 Short and Medium Term Storage

Facilities available at individual institutions and organisations may be used to store seeds, tubers, etc, for a few (1 to 5) years. Thereafter the entire collection may be grown, revived and restored.

7.3 Long Term Storage

Now the long term storage facility called National Gene Bank- which maintains –20C at 5 per cent grain moisture at NBPGR is constructed (Joshi and Hardas, 1976) and germplasm collections are stored for a much longer period (Table 4.1.).

Table 4.1 Regeneration period of seeds stored at –20C at 5 % moisture (after Roberts and Ellis, 1977)

Seed of crop plant	Regeneration after years
Rice	300
Wheat	78
Pea	1090
Onion	28

This would help save money and labour spent on annual regeneration and also the risk of mixing or losing some accessions (Roberts, 1972).

7.4 Live Plants

Perennial plants, trees, bulbous plants, etc, may be grown and maintained in a growing condition. This is space consuming but becomes a necessity unless cryopreservation is used. For example mango germplasm can be conserved by growing but also can be conserved in test tube by tissue culture or cryopreservation.

7.5 Tissue Culture

Tissue culture has been suggested as an aid to maintain germplasm of many problematic plants. Tissue culture form could as callus or small plantlets and stored under low temperature and in living form but growing slowly. Tissue cultured materials occupy less space to store. At the predecided intervals of time, these collections should be sub-cultured and stored again.

7.6   Cryopreservation

The principle behind cryopreservation is to bring the plant cell or tissue to a non-dividing or zero metabolism state by subjecting them to super low temperature. These super cooled tissues may be revived by rapid thawing at 40C in a water bath and then planted for revival depending on the tissue involved in preserving.

7.7   Natural Sanctuaries

This is also called *in situ* conservation. This type of conservation aims at conserving the land races along with their wild relatives in the same area. Some geographic areas are marked as gene reservoirs or gene sanctuaries or natural sanctuaries where some species can survive and maintain themselves, under natural conditions. For example, Garo hills of Meghalaya for Citrus, Khasi hills for Pitcher plant and Western Himalaya for *Dioscoria* have been marked a sanctuaries.

8.   ACHIEVEMENT AND USES OF INTRODUCTION IN INDIA

Introduced material may be utilised in any of the ways depicted in Figure 4.4. Some uses of introductions in India are summarised in the following page.

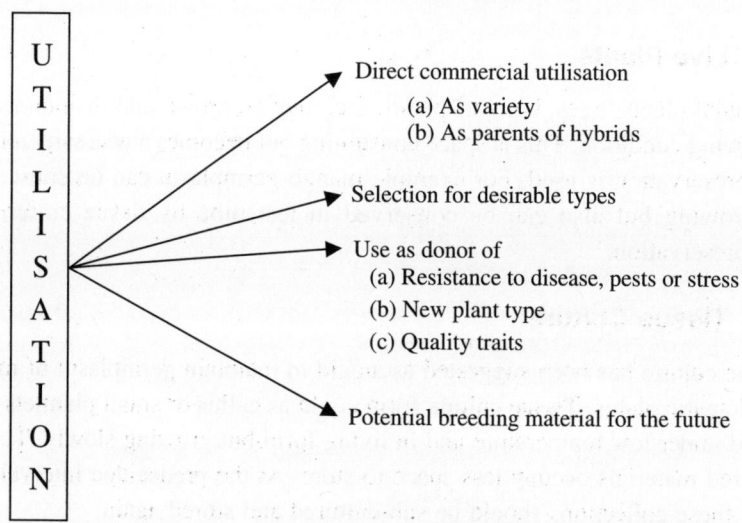

U
T
I
L
I
S
A
T
I
O
N

Direct commercial utilisation
    (a) As variety
    (b) As parents of hybrids

Selection for desirable types

Use as donor of
    (a) Resistance to disease, pests or stress
    (b) New plant type
    (c) Quality traits

Potential breeding material for the future

Figure 4.4   Utilization of introduced germplasm collections

8.1 New Crops

New crops like soybean, sugarbeet, sunflower, *Triticale*, Gobhi Sarson (*Brassica napus*), Karan Sarson (*Brassica carinata*) have been introduced in the recent past and are being grown in various areas. Hardee and Bragg varieties of soybean, Ramnos-Kaya of sugarbeet are direct introduction for cultivation. Armadillo and Badger of *Triticale* are varieties introduced with the introduction of new crops.

8.2 New Crop Varieties

(i) **Direct multiplication and release:**

If the introduced material is of known superior performance, it is multiplied and released as a variety. IR 8, IR 20, IR 26, IR36, IR42 and other IR varieties of rice from IRRI Philippines and many wheat varieties sonora 64 from CIMMYT Mexico were introduced and released as varieties.

A West Bengal collection of Lady's finger (*Abelmoschus esculentus*) resistant to yellow mosaic virus resulted in a well known variety, 'Pusa Sawni', which has been a ruling variety for the last 50 years. Bonneville and Arkel varieties of pea from, USA, Kent variety of Oat from Australia, Beauty seedless grapes from USA, and Tropical Beauty apple from Australia are some other examples of introductions.

Most ornamental flowers and plants, which beautify gardens and homes are introductions from other countries. Medicinal plants like *Solanum aviculare* and S. laciniatum, whose alkaloid is used in the preparation of oral pills for family planning, are introductions from New Zealand.

Some industrial crops and crop varieties like Chicory in Himachal Pradesh, Hops in Kashmir, La Bonita tomatoes for ketchup, Harbhajan Singh Matar for canning type pea are also introductions.

(ii) **Selection of desirable types:**

If the introduced material is a mixture or heterozygous, selection of superior types from it may be done in the next generation. Like wheat varieties, Sonalika, RR 21 and S 227 were selected independently from the same lot of wheat material introduced from CIMMYT, Mexico.

(iii) **Using as donor:**

If the introduced material does not prove superior agronomically to existing varieties, it may be used as donor for disease resistance, pest resistance, other stress resistance, for plant type and quality characters, etc, in hybridisation. Tetep and Tadukan donors of blast resistance in rice were introduced from Japan. Tift 23 A male sterile parent of hybrid bajra varieties (HB 1 to HB 5) was introduced from USA. In this connection it may be noted that the dwarfing source DGWG in rice and Norin in wheat, used in all dwarf varieties of today, are introductions from Taiwan and Japan respectively through other countries.

(iv) **Mutation Breeding:**

If the introduced material is agronomically superior but lacks a few characters, it is treated with mutagen to rectify the defect. This is a very quick method of developing a variety. For example, Sonora 64 variety introduced from Mexico had red grain and hence was not acceptable to farmers. An amber colour mutant, 'Sharbati Sonora', produced by Dr. M. S. Swaminathan with gamma-ray treatment became acceptable.

(v) **Potential Breeding material for Future:**

Plant introductions are stored or maintained in viable conditions to conserve and preserve their genetic variability for future use. The number of germplasm collected and being maintained at various institutions in India as of 1978 are as follows (Anon. 1978): Rice (43000), Bajra (3317), Sorghum (1600), Foxtail millet (1453), Kodo (824), Ragi (2719), Barnyard millet (559), Cotton (14,434), Jute (320), Linseed (3003), Sesame (3841), Urd (2810), Tomato (1549) Brinjal (942), Chillies (1370), Cabbage (328) and Cauliflower (661).

9. DISADVANTAGES OF INTRODUCTIONS

Though it has an overwhelming number of advantages, a few but serious disadvantages are also associated with introduction. Earlier there was no plant quarantine regulation in force, and hence these disadvantages were more common.

9.1 Disease

(i) Late blight of potato caused by *Phytophthora infestans*, got introduced in India in 1883 along with some potato accession and now it is the most dangerous disease of potato.

(ii) Bunchy top of banana, introduced in 1940, is causing serious losses.

(iii) Fire blight of apple and pear, caused by *Erwinia amylovora*, got introduced from England in 1940 and is now a problem disease in the northern hills.

(iv) Leaf disease of coffee caused by *Hemileia vastatrix* came to India in 1876 from Sri Lanka and is causing serious losses.

9.2 Pests

(i) Potato tuber moth entered in India in 1900 from Italy and is widely distributed in the country.

(ii) Fluted scale became a serious pest of citrus after 1928 when it entered India from Australia probably through Sri Lanka.

(iii) Wooly aphis, a serious pest of apple in North India now, is also an introduced one along with some apple accession.

9.3 Weeds

(i) Prickly poppy (*Argemone mexicana*) became a serious and problem weed, which entered the country through some foreign accession.

(ii) *Lantana camara* was introduced from Australia by the Britishers for ornamental purpose in India. Now this is the most problem bushy weed of the northern hills which not only poisons grazing cattle but also has replaced grazeable grasses from the hills. Singly, this weed has done maximum damage to cattle and forest.

(iii) *Phalaris minor* has become the major problematic weed of wheat after getting introduced from the USA. Weeding is not possible in the beginning due to its morphologic similarity with wheat, and later rouging is impossible. Thus farmers have no way but to spray costly herbicides to control it year after year..

10. USEFUL REFERENCES

Anon, 1978. Proc. Nat. Symp. On Plant and Animal Genetic Resources, New Delhi: IARI.

Chopra, V. L. (Ed) 2001. *Breeding Field Crops*. Oxford & IBH Publ. Co. Pvt. Ltd. New Delhi.

Bennet, E. 1970. Tactics of plant exploration. Genetic Resources in plant - their Exploration and Conservation. IBP Hand Book No.. 11. Blackwell Sci. Pub., Oxford.

Frankel, O. H. and Bennet, E. (eds.). 1970. Genetic Resources in Plants: Their Exploration and Conservation. IBP Hand Book No. 11. Blackwell Sci. Pub., Oxford.

Frankel, O. H. and Hawkes, H. G. (eds.) 1974. Crop Genetic Resources for Today and Tomorrow. Cambridge University Press, London.

Hill, A. F. 1976. *Economic Botany*. Tata McGraw-Hill, New Delhi, 560 pp.

Hutchinson, J.B. (ED.). 1965. Essays on Crop Plant Evolution. London: Cambridge Univ. Press.

Joshi, B. S. and Hardas, M. W. 1976. The proposed seed bank for plant genetic resources in India. Seed Research 4: 167-173.

Roberts, E.H. (ed.). 1972. *Viability of Seeds*. Chapman and Hall, London.

Robert, E.H. and Ellis, R.H. 1977. Prediction of seed longevity at subzero temperatures and genetic resources conservation. Nature 268: 432-433.

Simmonds, N. W. 1976. *Evolution of Crop Plants*. Longman, London.

Singh, H.B., Arora, R.K. and Hardas, M. W. 1975. Untapped plant resources. Proc. Nat. Acad. Sci. 41(3): 194-203.

Smartt, J. 1976. *Tropical Pulses*. Longman, London, 348 pp.

Vavilov, N.I. 1951. The origin of variation, immunity and breeding of cultivated plants. Chronica Bot. 13: 1-364.

Whyte, R. O. 1958. Plant Exploration, Collection and Introduction. F.A.O. Agricultural Studies No. 41.

Zeven, A. C. and Zhukovasky, P.N. 1975. Dictionary of Cultivated Plants and Their Centres of Diversity. Wageningen: PUDOC.

Breeding Self-pollinated Crops by Introduction and Selection

The breeding methods enumerated in Chapter 3, which are used in breeding naturally self-pollinated and often self-pollinated crops, are discussed below.

1. INTRODUCTION

If the purpose of all breeding efforts is to make the best varieties of crops available, no method can be quicker than plant introduction. Since plant breeding programmes are global in nature and every country has her own programme the best variety developed anywhere can be introduced, acclimatised and cultivated quickly. Many such examples are cited in Chapter 4.

In India we have been fortunate to have so many native crops. But in recent years new crops like soybean and triticale have been introduced from the USA and Hungary respectively. New crop varieties have been the major contributions of introduction as a method of breeding. Lerma Rojo and Sonora 64, introduced in 1964, was the result of over 10 years' breeding effort of Nobel laureate Dr. Norman E. Borlaug in Mexico. Rice variety Taichung (Native) 1 bred in Taiwan in the 1950s could change the concept of rice farmers and breeders in India on being introduced here in 1965. Had it been introduced 15 years earlier, there would have been a major breakthrough in rice production much earlier.

The existing variability of crops is evolving further under the influence of man and nature. All these variability are collected continuously, and their areas are revisited at pre-decided time schedule. Thus plant introduction would always have active role to play as a method of breeding.

2. SELECTION

In its simplest form, selection, that is, choosing plants of one's choice, must have been the first step in breeding crop plants by the first agriculturists. Thus choosing the best out of one's crop, continued over generations, has played an important part in the development and retention of developed varieties.

2.1 Principle of Selection

The genetic diversity providing the basis of selection has its origin in spontaneous mutations and its subsequent incorporation into population through natural hybridisation and recombination over generations.

In self-pollinated crops and crops with very low degree of cross-pollination, individuals in a population are homozygous. Whatever variability exists for various plant characters between individual plants of a population, can be fixed in one cycle of selection. For example, selecting a dwarf plant out of a population can be completed in one cycle and no further segregation would take place. Under these conditions further selection can be done only when variation has been created through hybridisation or mutation.

2.2 Types of Selection

Selection in self-pollinated crops is performed in two groups of populations. The first is in the bulk population from farmers' fields where mass selection and pure line selection can be done. The second is selection from the segregating material of a cross by pedigree method, bulk method or single seed descent method, which are discussed in Chapter 6. In natural populations or populations from farmers' fields, any of the two methods of selection, namely mass selection and pure line selection, are followed.

2.2.1 *Mass selection*

This is one of the oldest methods of breeding crop plants. The procedure is very simple. The best plants out of the field or bulk are selected and threshed together. The resulting bulk harvest is used for raising the next crop generation. Mass selection is practised in mixed population of cultivars or land variety or unimproved strains. Land variety is a well adapted local cultivar (cultivated variety) which was release many years back. Such varieties become a mixture of many minor variant forms, although possessing similarities for major agronomic features. Thus mass selection is the process of selecting the best and bulking the seeds till desired result is achieved. The unit of selection may be a plant, earhead, individual seed or fruit.

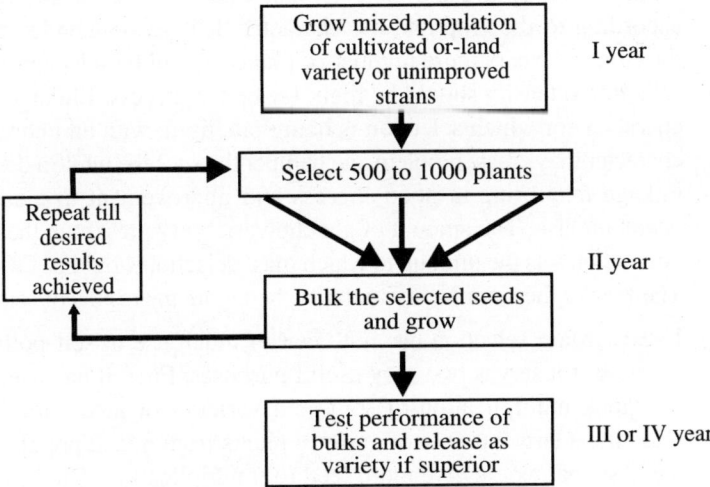

Figure 5.1   Procedure of mass selection

(i) **Procedure of selection:**   Plants are selected on the basis of phenotype of any character, etc. Plants may be selected either from the whole field or the field may be divided into several sub-plots or compartments. Selection of equal number of plants may then be made from each compartment. The procedure depends on the type of crop and the character to be improved, but a general procedure has been outlined in Figure 5.1. Generally, 500 to 1000 plants are selected and bulked. Then the selected bulk is tested for performance against the most popular cultivar, and local check. If the bulk is found superior, it may be released as a variety.

(ii) **Success:**   The selections of plants, earheads or fruits are made on the basis of appearance, that is, phenotype. Therefore the success of mass selection would depend on heritability, size of population, intensity of selection, linkage relations and variability of the characters. Heritability would determine if the performance of selected individual is going to be the same or different in the next generations. If heritability is high, the selected individuals would have the same good expression of the character. If heritability is low, a vigorous plant may produce the poorest progeny and vice versa, and therefore mass selection would not be useful.

(iii) **Rule:**   There is no rule about the size of population grown each year and the number of individuals selected out of it (selection intensity). It may vary from 2 per cent (2 plants out of 100) to 20 per cent according to the crop and the character. It is recommended that in the first few years more number of plants should be selected, that is, selection intensity should be high, for better success. Linkage of the character for which selection is being practised, with an undesirable character may create problem and hamper the success, while a desirable linkage may bring in good success and improvement in the linked character also. The amount of genetic variability present in the initial population sets the limit up to which mass selection would be effective. The greater the amount of variability, better are the results of selection.

(iv) **Uses:**   Mass selection has not been of much use in self-pollinated crops. But it serves two very useful purposes. First, it has been used for quick multiplication of some old varieties or land varieties, as selecting a large number of uniform plants from mixed population in the first year would give more seed for multiplication. Thus a variety can be purified and multiplied quickly, which is not so in pure line selection. Earlier, some varieties like TMV1 and TMV2 of groundnut were developed in Madras using mass selection in Saloum and Spanish varieties, respectively. The second use to which mass selection is put in a modified way is the production of Breeder's Seed. A few hundred individual progenies are selected from a field and grown the following year in individual rows and observed for critical characters. Selected progenies are bulk harvested as breeder's seed, which is used for the production of Foundation Seed.

(v) **Drawback:**   Mass selection has some disadvantages. First, it tends to be a relatively slow method for changing the desirable gene and genotypic frequencies when compared to the pure line method. Second, mass selection only increases the frequency of genotypes already present in the population, but does not provide opportunity for recombination.

2.2.2   *Pureline selection*

As the name indicates, this type of selection is to purify on line basis and to develop the pure lines which breed true. The term pure line was suggested for the first time by a Danish botanist, W.L. Johannsen (1903) while working with the Princess variety of French bean (*Phaseolus vulgaris*). Pure line may be defined as a strain obtained from the progeny of a single self fertilised homozygous individual. In his detailed work, Johannsen took a commercial sample of the Princess variety of beans. From this sample he selected large and small seeds and planted them. He harvested the seeds from individual lines and found striking differences in the seed weight of these lines. Further, he selected 19 plants representing extreme seed weights- from the lightest to the heaviest.

Table 5.1   Mean seed weight of some of Johannsen's pure lines in Princess beans

Pure line No.	Mean seed weight (mg)	Range (mg)
1	642	649 – 631
10	465	469 – 421
19	351	358 – 348

For six successive generations the smallest seeds of pure line, for example line 19, were found to grow into plants with seeds of the same average weight (351 mg). Similarly, the largest beans selected for six successive generations from pure line 1 were found to grow into plants with seeds of 642 mg. Variations in seed size which existed in individual lines were due to position of pod on plant or position of seed in pod. That is, 649 mg or 631 mg seeds of line 1 have the same genotype and the difference in their weight was due to non-hereditary or environmental reasons. On sowing both the seeds gave progenies with the same average seed weight, that is, 642 mg. Thus line 1 became a pure line and further selections in it for seed size were not effective. Johannsen concluded that differences in average seed weight of different lines are hereditary but between the individuals of a line are non-hereditary.

Since French bean is a naturally self-pollinated crop, the sample of the princess variety must have been a heterogeneous mixture of 19 homozygous lines. By one cycle of selection, these types were separated into 19 lines. But once the homozygous lines were isolated, the differences within the lines were only environmental and thus further selection was ineffective.

(i) **Procedure of pureline selection:**   There are three distinct steps in pureline selection. The first is the selection of large number of individual plants from the population. There is no fixed rule for the number of plants, but initially it should be as large as possible. An average range of 200 to 1000 plants may be kept. The procedure has been outlined in Figure 5.2.

The second step consists of growing progeny rows from individual plants for visual observations. This 'eye-ball' selection and rejection continues for a few years. No selection is practised within line (progeny). Lines with certain defects or susceptible to diseases, etc, are eliminated. A drastic reduction is made in the number of lines during this process. The third step consists of evaluation of selected lines is replicated trials for 2-3 years. The superior-most line is selected, multiplied and released as variety.

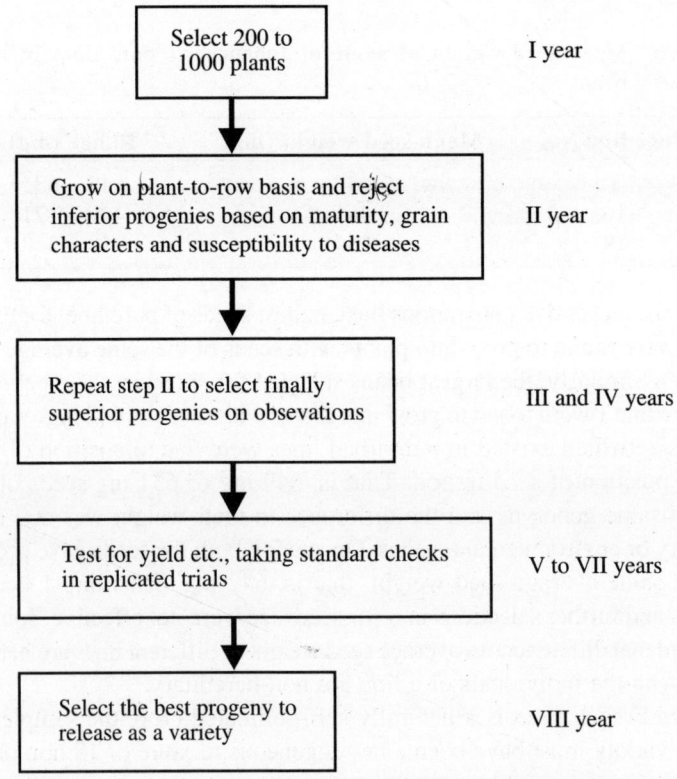

Figure 5.2   Procedure of pureline selection

(ii) **Uses:**   Pure lines have a special significance in the improvement of self-pollinated crops. Mixed populations, varieties from the farmers' fields and unimproved varieties may be used to isolate superior pure lines and released as varieties. Many varieties have been bred using this method. The majority of so called improved varieties of rice, wheat, pulses, oilseeds, etc, have been bred using pure line selection. In the past the pure line method of breeding was as popular as the pedigree method now. For example, NP 4 and NP 6 were the predominant wheat varieties. Prior to 1960, about 430 rice varieties were released, out of which around 400 were bred using pure line selection. Basmati 370 and T 3 are still the finest type of rice varieties which were bred using the pure line method. Kalanamak KN 3 variety of aromatic rice was released in U.P. during 2010 by pureline selection from a landrace Kalanamak, which had deteriorated for aroma and grain quality. T 9, RT 11 and Laha 101 of rai (B. juncea) bred long back still continue in cultivation. Similar examples are there in all self-pollinated crops.

Limitations

1. It cannot be used in cross-pollinated crops for development of varieties with the same ease and success as in self-pollinated crops. Selfing or sib mating has to be practised every year, which is laborious.
2. In cross-pollinated crops like carrot, etc, where inbreeding depression is so high that inbreds cannot be developed.
3. No new genotypes are created by pure line selection. Improvements by this method are thus limited to isolation of the best genotypes already existing in the population.

Various points in mass selection and pure line selection have been outlined in Table 5.2.

Table 5.2   Comparison of mass and pure line selection

Mass selection	Pure line selection
1. Used mostly in cross-pollinated crops, in self pollinated it has limited use.	Used usually in self-pollinated crops, in cross it has limited use.
2. Comparatively larger number of plants are selected in the first year.	Comparatively lesser.
3. Selected plant, seeds or earheads are bulked to grow the crop in subsequent years.	Not bulked, but the seed of each plant is grown separately on plant to-row basis in subsequent years.

4. No separate progeny rows and hence testing of bulked progenies not done.	Performance of each selected progeny is tested.
5. Several plants bulked to produce the variety.	Only one superior-most line multiplied to produce a variety.
6. The developed variety, though less heterogeneous than the original, yet is not homogeneous	Variety developed is homozygous and homogeneous.
7. Mass selection has to be continued to maintain the variety.	No need to continue pure line selection to maintain the variety.
8. Takes comparatively less time to develop a variety (about 8 years).	Takes little longer time (10 years).
9. New variety is a mixture of inbreds i.e. heterozygous plants and heterogeneous population.	New variety is a pureline i.e. homozygous plants and homogenous population

3. USEFUL REFERENCES

Allard, R.W. 1960. *Principles of Plant Breeding*. John Wiley & Sons Inc., New York

Chopra, V. L. 2000. *Plant Breeding: Theory and Practice*. Oxford & IBH, New Delhi.

Briggs, F. N. and Knowles, P.F. 1967. *Introduction to Plant Breeding*. Reinhold Publ. Corpn., New York.

Hayes, H.K., Immer, F.R and Smith, D.C. 1955. *Methods of Plant Breeding*. McGraw-Hill Book Co., New York.

Hoffman, W., Mudra, A and Plarre, W. 1970. Lehrbuch der Zuchtung Landwirtschaft-licher Kulturpflanzen. Berlin: Verlag Paul Parey.

Johannsen, W. L., 1903 Uber die Erblichkeit in Populationen und in rheinen Linien. Jena: Verlag Fischer.

Lerner, I. M. 1958. *The Genetic Basis of Selection*. London: Chapman and Hall Ltd.

Simmonds, N. W. 1979. *Principles of crop Improvement*. London: Longman.

Breeding Self-pollinated Crops by Hybridisation

The main object of hybridisation in breeding self-pollinated crops is to combine desirable characters that are found in two or more genotypes into one. The first step obviously is to cross these different genotypes.

After the cross has been made, F_1 (first filial generation) is grown and seeds set on it are harvested. This makes the seed source for raising F_2 (second filial generation) population. The F_2 population is the first segregating generation where selection is generally done. There are various procedures of making selection and purifying it: Pedigree method; Bulk method; Modified bulk method; Single seed descent method; and Back cross method.

1. PEDIGREE METHOD

After the cross has been made in selected parents, and F_1 plant has been grown, seeds produced on it are collected. On sowing, these seeds give rise to F_2 population where segregants appear. Selection of the desired type of plants starts from F_2 population and individual plants are carried through F_3, F_4, F_5, F_6 and so on generations till the segregation occurs. Selection is made within and between the lines. A proper kind of record is maintained for each selection, such that it is possible to trace each pedigree line to a specific plant of F_2 generation.

1.1 Procedure of Handling

The procedure as outlined in Figure 6.1 is explained as follows:

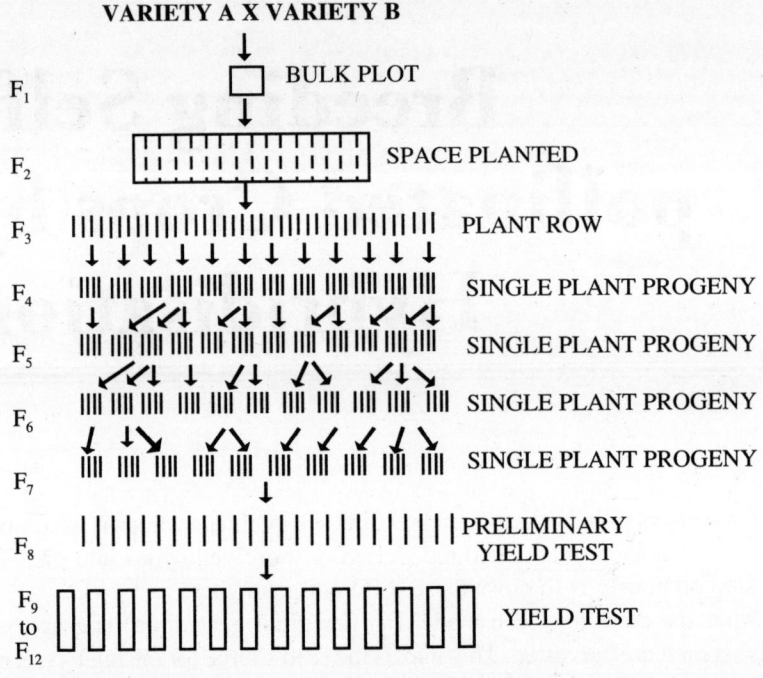

Figure 6.1 Procedure of Pedigree method of breeding.

1st year: Crossing of selected parents

2nd year: Raising of F_1 plants to produce enough F_2 seeds.

3rd year: Space planting of large F_2 population (2000-10,000 plants) and selection of a number of plants (200-500). To aid selection, diseases and pests may be artificially inoculated. Selected plants may be tagged any time between flowering stage till maturity. Checks of promising varieties should be grown after every 20th row to aid in selecting better segregants. Thus only resistant plants having desirable appearance may be selected. Each selected plant is harvested, threshed and kept separately.

4th year: Each plant selected in F_2 is grown in progeny rows in F_3 generation. About 20 to 100 plants are grown in each plant progeny separately under space planting. The F_3 population may be as large as F_2. Individual plants are again selected from promising progenies. These plants are harvested, threshed and kept separately.

5th year: The F_4 generation is handled in much the same ways as the F_3 generation but emphasis is placed on selection among progenies. Some progenies may appear fairly uniform, but several others may still segregate. A few (10) best plants may be selected from each of best progenies.

6th year and onwards: Normally, segregation does not continue beyond the F_6 generation. If any progeny segregates, it is rejected unless very promising. Uniform and promising progenies are selected and bulk harvested for evaluation in field trials.

Usually trials are initiated in F_6 or latest in F_7 and progenies are compared with the best varieties. If any progeny is found better than the existing varieties in repeated tests for three to five years at several locations, it is released as a variety.

Thus it takes about 12 to 14 years to develop a variety by the pedigree method. But with the help of a glass house two generations, such as crossing and F_1, can be grown in one year. Still, with the help of 'off-season nursery', every year two generations can be grown and time taken can be shortened from 12 years to 6 years. For example, after the main crop of wheat is over in March, an off-season crop from April to September is grown in Wellington (Tamil Nadu). In rice, after the main season is over in November, an off-season crop is grown in the south (Cuttack or Hyderabad) from December till May.

After the rediscovery of Mendelian Laws of inheritance, the pedigree method has been the most predominant in breeding self-pollinated crops. It is quick, precise and well suited to those crops where individual plants can be harvested easily. Today almost every variety developed in self-pollinated crops owes its development through the pedigree method. It has some drawbacks, such as it involves a large amount of work in early generations, uncertainly of success of selection for complex characters, and limited population and material that can be handled.

2. BULK METHOD

This method of breeding differs from the pedigree method in that no selection is practised in F_2 to F_5 generations. By the F_6 generation a high proportion of plants will become homozygous. This method is also known as the mass or population method. Bulk plots of F_2 to F_6 may be subjected to disease, pests and other stresses to eliminate the susceptible genotypes. Natural selection would eliminate the weaker types. F_2 is harvested and threshed in bulk, and F_3 is grown by taking a random sample from the bulk produce of F_2. This procedure is repeated till F_6 and then single plant selections are made. These plants are grown into single plant progenies in the following generation. The best progenies are selected based on promise and uniformity. These are tested in yield trials and the most promising one is released as a variety (Figure 6.2). Nilsson-Ehle of Sweden was first to use the bulk method and it is in use ever since. The bulk method is simple, convenient, inexpensive and less labour consuming, but it takes more time. The other drawback of the method is its total dependence on natural selection to select the superior types. The more

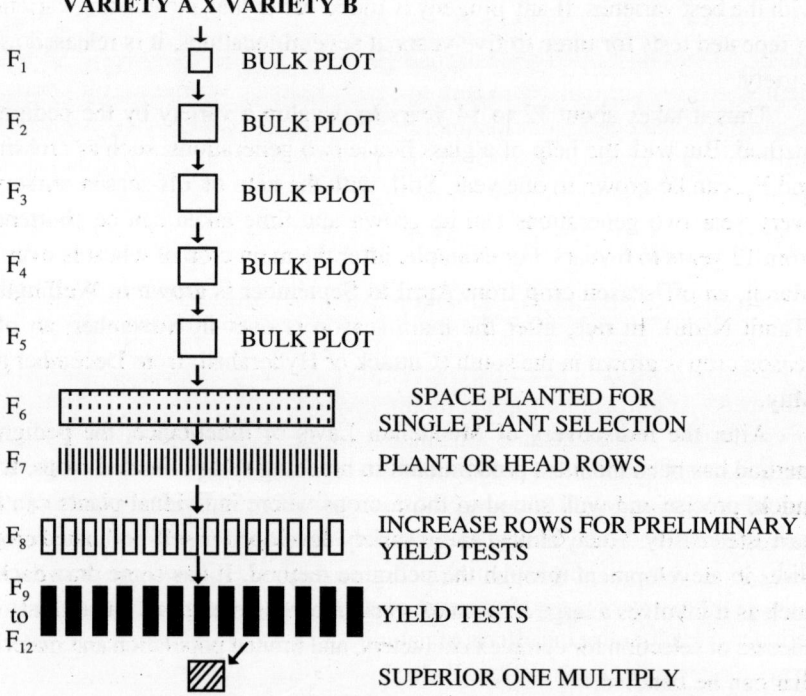

Figure 6.2   Procedure of Bulk Method of breeding.

competent or surviving types, which get selected in this method, may not necessarily be the best yielding types. To overcome this drawback and to incorporate the plus points of the pedigree method, several modified bulk methods have been proposed. A comparison of the pedigree and bulk methods has been given in Table 6.1.

3. MODIFIED BULK METHODS

The modified bulk methods incorporate one or a few steps of the pedigree method and hence should be classed as intermediates between the two.

One such modification is when plants are selected right from F_2 but the produce of these plants is bulked to grow the F_3 generation. Single plants are again selected and bulked to grow the F_4 generation. This selection and bulking is done up to F_6 and then selected plant are kept separately and grown in progeny rows. The best progenies are then selected and tested in yield trials. During F_2 to F_6 stage, the bulk plots may be subjected to diseases, pests and other stress conditions depending on breeding aims. Harington (1937) used a ' mass-pedigree' method in which bulking can terminate when the season is favourable for single plant selection. Bulking may stop any time between F_2 and F_6. Single plant selection thus made may be carried further by the pedigree method. This is particularly useful, for, while breeding for disease and insect resistance, if the epiphytotics cannot be created in a particular year, it may be bulked. Single plant selection may be made in other years. Similarly, there are many modifications based on the time of individual plant selection and testing, etc.

4. SINGLE SEED DESCENT METHOD

In principle the Single Seed Descent (SSD) method is a modification of the bulk method of breeding. But the modification is such that it allows the equal survival of all segregants and thus merits special mention. This method was first suggested by Goulden (1941) but Grafius (1965) and Brim (1966) refined and advocated it. With this procedure one or two seeds are collected from each F_2 plant and then bulked to grow the F_3. Similar practice is continued till F_5 or F_6 or when single plant selections are made and evaluations done. It is designed to preserve the total range of variation throughout the propagation period and to minimise the effects of natural selection in changing the genotypic array in the original population. Ikehashi (1977) proposed a modification of SSD specially for advancing generations of rice crosses, which involve photosensitive parents. In normal course, such crosses take 160 days or more to mature and only one generation can be grown per year. But by the new technique, called Rapid Generation Advance (RGA) technique, early generations (F_2 to F_8) are planted at close spacing (1000 plants/sq m as against 35 plants of normal) in a green house or phytotron under short days and high temperature. This shortens the crop maturity to 100 days by reducing thermo and photo dependent periods and thus three generations can be grown per year. Every plant produces a few grains, which are carried to the next generation. There is no danger of less competitive segregants being lost in any generation. Single plant selection may be made by growing F_6 in normal field conditions. RGA has an additional advantage in cases where the improvement sought is governed by recessive genes or in unfavourable linkages. By F_6 these become homozygous and linkages break if at all there is any chance.

Table 6.1 Comparison of pedigree and bulk methods of breeding

Pedigree method	Bulk method
1. Desired plants are selected in F_2	No selection is made in F_2
2. Single plant progenies are grown after making selection each generation till material becomes homozygous.	Bulk is grown in each generation up till F_6 and then single plant selections are made.
3. Pedigree of each progeny is maintained so that even the finally selected progeny can be traced back to a particular F_2 plant	No pedigree is kept and hence it is not possible.
4. Artificial selection is practised, which permits the plant breeder to exercise his skill to the maximum.	Only natural selection operates and skill of the plant breeder does not play any role.
5. There is a tremendous rejection of undesired segregants in F_2.	No rejection but almost all the segregants are carried forward till F_6 generation.
6. Laborious and expensive.	Less laborious and inexpensive, thus large number of crosses can be handled.
7. Most popular method of breeding.	Not so popular.

5. BACK CROSS METHOD

Back cross is the crossing of F_1 with either of the parent. Test cross is also a back cross of F_1 but with the recessive parent. Harlan and Pope (1922) proposed back cross as a method of breeding. It is employed in the improvement of both self- and cross-pollinated crops, where a variety is deficient in one or two aspects. Thus, one parent is an established variety called recurrent parent and the other one, donor or non-recurrent parent, from where the desired genes are to be transferred. The following four requirements should be met for a successful back crossing programme: (1) a satisfactory recurrent parent must exist, (2) high expressivity of the gene being added, (3) simple testing technique for detecting the added gene, (4) recovery of the recurrent genotype in a reasonable number of generations.

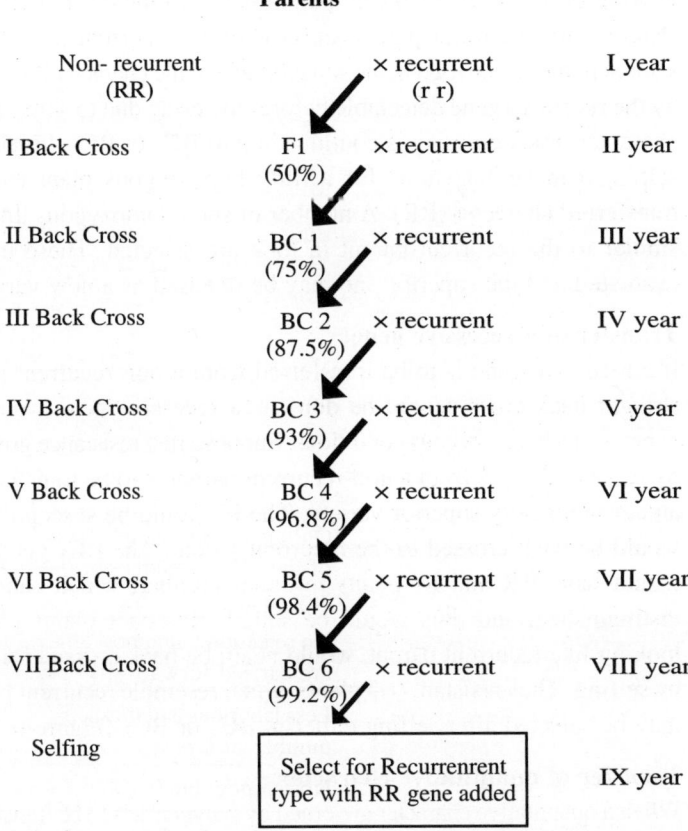

Parents

Non- recurrent (RR)	× recurrent (r r)	I year
I Back Cross	F1 (50%) × recurrent	II year
II Back Cross	BC 1 (75%) × recurrent	III year
III Back Cross	BC 2 (87.5%) × recurrent	IV year
IV Back Cross	BC 3 (93%) × recurrent	V year
V Back Cross	BC 4 (96.8%) × recurrent	VI year
VI Back Cross	BC 5 (98.4%) × recurrent	VII year
VII Back Cross	BC 6 (99.2%) × recurrent	VIII year
Selfing	Select for Recurrenrent type with RR gene added	IX year

Figure 6.3 Procedure of Back Cross method of breeding to transfer a dominant gene (RR). Number in parentheses indicate the percent of genotype of recurrent parent.

5.1 Procedure

Depending on the situation there are four schemes of back crossing. The number of back crosses varies from 3 to 10 but it should be continued only till desired characters of recurrent parents are transferred.

(i) **Transfer of a dominant gene:**

In the simplest form a back cross programme consists of crossing recurrent and non-recurrent parents and thereafter repeated crossing of selected plants from F_1 and later back cross (BC_1 to BC_5) generations to recurrent parents (Figure 6.3). In each back cross generation, plants looking closer to the recurrent parent but with the desired character (RR) of non-recurrent parent are crossed to recurrent parents. The same scheme can be used in the situation where the character is governed by the recessive gene detectable before flowering due to some marker character. Back crossing is continued up to BC_5 or BC_6. Finally it is selfed to make selections for a fully homozygous plant even for transferred character (RR). A number of such homozygous lines and similar to the recurrent parent in look are selected. These may be evaluated and the superior one may be released as a new variety.

(ii) **Transfer of a recessive gene:**

If a recessive gene is to be transferred from a non-recurrent parent, straight back cross cannot be done as a recessive gene would not express in a heterozygous conditions. Suppose rust resistance governed by recessive genes from a non-recurrent parent is to be transferred to an agronomically superior variety. The F_1 would be susceptible but would be back crossed to the recurrent parent. The BC_1 generation would have RR and Rr plants in equal number, which cannot be distinguished, and thus would be selfed. Resistant plants (*rr*) but looking like recurrent parent, would again be back crossed followed by selfing. Thus resistant (*rr*) plants which resemble recurrent parents may be selected after selfing either in BC_3 or BC_4 (Figure 6.4).

(iii) **Transfer of quantitative character:**

When a quantitative character governed by polygenes is to be transferred, each back cross generation should be selfed and F_2 and F_3 generations should be grown. If the character has low heritability, back cross F_2 and F_3 populations will have to be large (Briggs and Knowles, 1967). Back crosses are made on plants selected in F_3, and BC_2 and BC_3

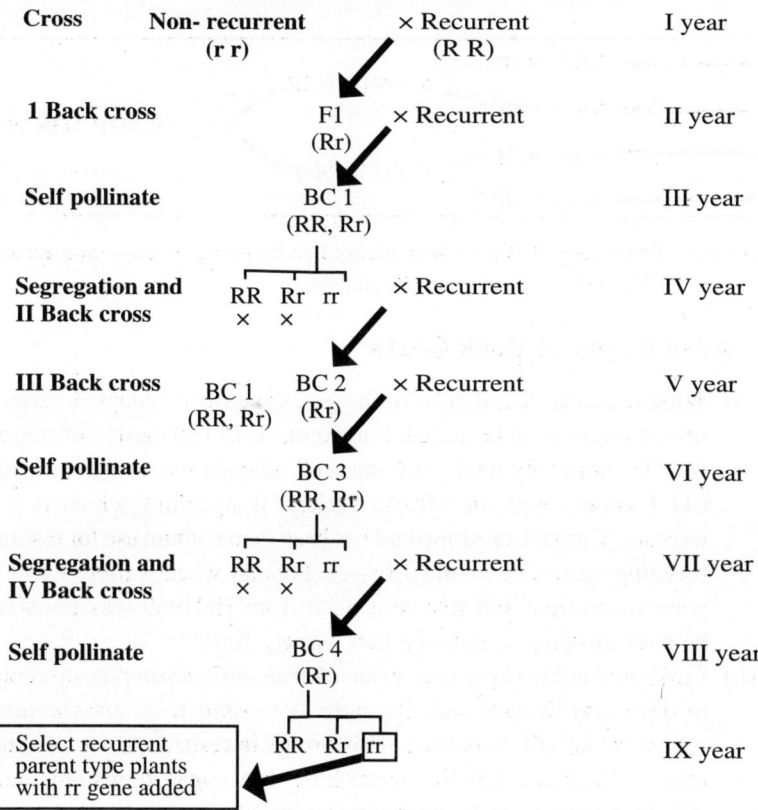

Figure 6.4   Procedure of Back Cross method of breeding to transfer a recessive gene (rr). In this case back crossing and selfing alternates.

generations are grown. The system of back crossing and selfing alternates till the desired results are obtained.

(iv) **Simultaneous use of more than one non-recurrent parent:**
Mac Key (1954) suggested a modified back cross programme where up to eight non-recurrent parents could be utilised first in multiple crossing and then in back crossing to recurrent parent in the usual manner. Use of four non-recurrent parent (N) and one recurrent (R) parent has been shown in Figure 6.5. Selections made from III cycle of crossing may be further back crossed to recurrent parent, to initiate a back cross programme with a recurrent (R) parent.

Parents	Cycle of crossing

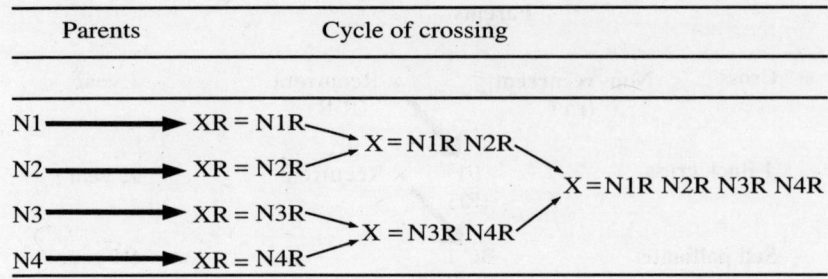

Figure 6.5   Procedure of Back Cross method of breeding to use 4 non-recurrent (N) and one recurrent (R) parent.

5.2   Advantages of Back Cross

(i)   Without much disturbance in the genotype of the adapted variety, a few characters can be added. For example, CO 20 variety of sorghum was developed by back crossing well adapted but striga susceptible CO 1 variety with an African variety 'Bonganhilo' which is striga resistant. The back cross method has been in maximum use for resistance breeding against pests and diseases. Popular wheat variety K852 had gone susceptible and rust resistance from HD1969 was transferred by back crossing to develop new variety K8027.

(ii)   Cultivated noble sugar canes (*Saccharum officinarum*) is susceptible to pests and disease and therefore is crossed to *S. spontanium* to obtain the latter's resistance. This brings in resistance but also many undesirable characters like more fibre, low sugar, thin stem, etc. by two back crosses of F_1 with noble cane, these undesired characters are removed. This is known as nobilisation of canes.

(iii)   Quality in rice is a complex character and is difficult to be recovered through the normal pedigree method of breeding. a variety, Sabarmati, was developed by five back crosses of F_1 (Taichung Native 1 x Basmati 370) with Basmati 370. Sabarmati is a semi-dwarf, scented and high yielding variety.

(iv)   Back cross breeding is independent of environment and hence can be used in any environment. More than one generation can be grown to make crosses even in off-season nurseries.

(v)   Adaptability of the developed variety is not a problem. It does not need any critical test for performance and characters, as the recurrent parent is well-adapted variety.

(vi)   It takes lesser time to develop a variety.

(vii)   Small populations of plants are needed in each generation.

(viii) If character to be transferred is linked with an undesirable one, the chances of recovering a cross over genotype is more in this method compared to the pedigree method. This is due to the fact that in back cross heterozygosity is continuous for character under transfer.

(ix) Component lines may be developed to form multilines.

5.3 Limitations

(i) Not suitable where crossing is difficult, as the crossing is to be done in every generation.

(ii) It does not permit the unusual combination of genes from two or more varieties as expected in usual hybridisation.

(iii) Characters of low heritability are hard to transfer.

(iv) It only stabilises production but does not increases it as the developed variety is not very different from the old one in yield potential. Various schemes of back crossing used under different situations have been compared in Table 6.2.

6. TYPES OF VARIETIES AND THEIR DEVELOPMENT

The following types of varieties are normally used in self-pollinated crops:

1. Homozygous, homogeneous / purebreds / purelines
2. Hybrids
3. Multilines
4. Multilineal hybrids

6.1 Purebreds / Purelines

Such types of varieties are homozygous, homogeneous and breed true. These are developed using any of the methods described for self-pollinated crops. Most varieties of self-pollinated crops belong to this category (see Chapter 14).

6.2 Hybrids

The varieties are hybrids and are heterozygous but homogeneous and do not breed true. For example, hybrid varieties of rice (see Chapter 9), maize, sorghum, cotton, brinjal (Vijay) and chillies (Chamatkar) are under cultivation. But every year fresh seed has to be produced and used. The methods to develop such varieties have been described in detail in Chapter 9.

Table 6.1 Comparison of pedigree and back cross methods of breeding.

	Pedigree method	Back Cross method
1.	Desired plants are selected in F_2 and subsequent generations for purification.	Continuous selection for the desired segregants is made and crossed with the recurrent parent.
2.	Single plant progenies are grown after making selection each generation till material becomes homozygous.	Crosses of the selected plants are made with the recurrent parent till its desired genotype is recovered.
3.	Pedigree of each progeny is maintained so that even the finally selected progeny can be traced back to a particular F_2 plant	No pedigree is kept and hence it is not possible to trace back.
4.	Artificial selection is practised, which permits the plant breeder to exercise his skill to the maximum.	Artificial and natural selection pressures are made to identify the desired segregants for crossing purpose. Skill of the plant breeder does play important role.
5.	Population is large and there is a tremendous rejection of undesired segregants in F_2.	Population in the back cross generations is small and only undesired plants are rejected.
6.	Laborious and expensive.	More laborious and expensive as putting artificial selection pressure and crossing with the recurrent parent is involved.
7.	Most popular method of breeding.	No so popular but under specific situation of improving a recurrent parent, there is no better method than this.
8.	New variety developed may or may not be similar to the original parents.	New variety is very similar to the recurrent parent except for the character transferred from the non-recurrent parent.
9.	New variety is expected to be much superior to the parental varieties.	New variety is not expected be superior to the recurrent variety in yield.

Table 6.2 Various schemes of back crossing suggested under the genetic control of the character under transfer.

	Either RR or pre-flowering detectable character	Polygenes	rr, not detectable	Many donors
I	Cross parents	Cross parents	Cross parents	Cross parents
II	F_1	F_1	F_1	Cross single crosses.
III	BC1 (7 plants)	BC1 (7 plants)	BC1 (large)	Cross double crosses
IV	BC2 (7 plants)	F_2	F_2 (large)	F_1
V	BC3 (7 plants)	BC2	F_3 (large)	BC1
VI	BC4 (7 plants)	BC3	BC2 (large)	BC2
VII	BC5 (7 plants)	F_2	BC3	BC3
VIII	BC6 (49 plants)	BC4	F_2	BC4
IX	F_2 (49 progenies)	BC5	F_3	BC5
X	F_3	F_2	BC4	BC6
XI	–	F_3	F_5	F_2
XII	–	F_3		F_3

6.3   Multiline

Multiline cultivar is a mixture of a number of back cross derivatives of a recurrent parent, each component of a multiline has uniform maturity, height, grains, etc, except having a different resistant gene. Jensen (1952) originally proposed the use of multiline cultivars in oats. Borlaug (1953) proposed a clear cut procedure to develop and use them to control rust disease of wheat. Details about their use to control disease have been described in Chapter 12. Usually 4-10 lines are bulked to make a multiline variety. Bithur was the first multiline wheat variety in India released in 1978 in Uttar Pradesh. Now two more multilines, KSML-3 and MLKS 11, have been released. All the three varieties are multilines of Kalyan Sona and have 9, 8 and 8 component lines respectively. The following methods may be used to develop multilines.

   (i) **Back cross method:**
        Borlaug (1953) suggested that through back cross different resistance genes may be transferred in different lines. After six back crosses selections may be practised for uniform lines. These lines may be composited to make multilines.

  (ii) **Restricted back cross method:**
        Borlaug (1959) realised that making six back crosses to develop component lines is difficult. Thus only three back crosses may be made and selection may be made for segregants with uniform agronomic features. Component lines so developed may be used to constitute multilines.

 (iii) **Segregants from multiple cross:**
        It was further simplified that in multiple crosses where many donors for resistance have been used, selection may be made for uniform type of segregants. On test if they are found to have different resistant genes, they may be composited to get multiline.

6.4   Multilineal Hybrids

Multilineal hybrids were also proposed by Borlaug (1965). He proposed that in multiline varieties, if in place of fixed component lines hybrids are utilised, it will form a multilineal hybrids variety. Thus the advantages of multilines as well as of heterosis would be utilised in such a variety. The procedure to develop a multilineal variety has been outlined in Figure 6.6. A and B parents are so chosen as to produce heterotic hybrids. One resistant gene and cytoplasmic male sterility is incorporated in A parent. Other six resistant genes are incorporated in A parent to prepare six component lines. The eighth resistant gene and fertility restorer is incorporated in B parent. First A (S) is crossed with other

six A lines to produce six male sterile F1. This is pollinated with B8 (F)
parent to produce six hybrids, which are mixed to produce a multilineal hybrid.

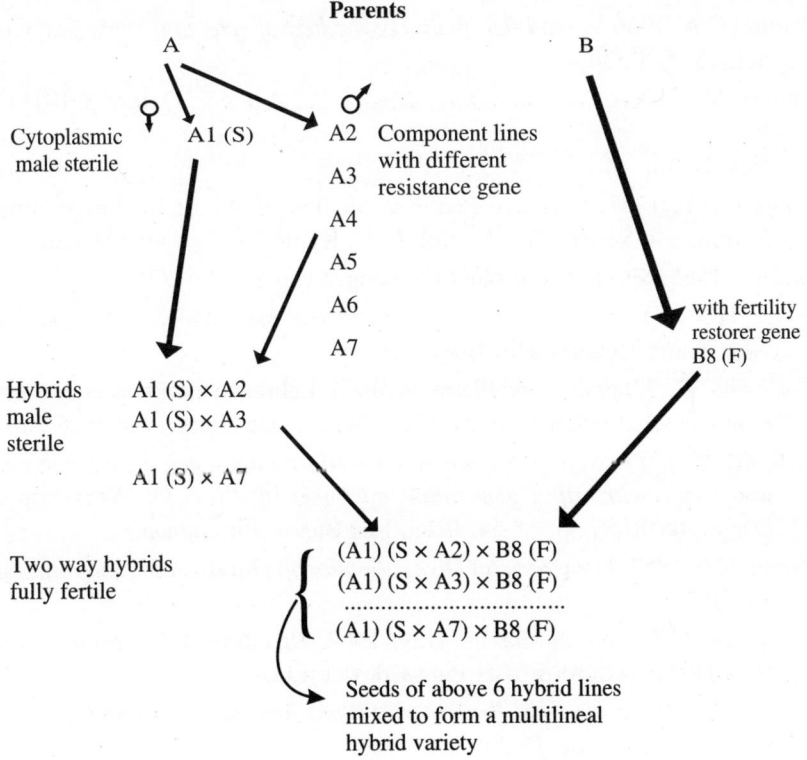

Figure 6.6   Procedure to develop a multilineal hybrid variety.

7. USEFUL REFERENCES

Allard, R.W. 1960. *Principles of Plant Breeding.* New York: John Wiley &
    Sons

Andrus, C.F. 1963. Plant breeding systems. Euphytica 12: 205-252.

Anon, 1976. Multilines, safety in numbers. CIMMYT Today No. 4, 11 pp.

Banga, S.S. and Banga, S.K. 1998. *Hybrid Cultivar Development.* Narosa
    Publ. House, New Delhi.

Borlaug, N.E. 1959. *Use of multilineal or composite varieties to control air
    borne epidemic diseases of self-pollinated plants.* Proc. I. Int. Wheat Genet.
    Symp., Winnipeg. Canada.

----. 1965. Wheat, rust and people. Phytopathology 55: 1088-1098.

Brings, F.N. and Knowles, P.F. 1967. *Introduction to Plant Breeding*. New York: Reinhold Publ. Corpn.

Brimm, C.A. 1966. *A modified pedigree method of selection soybeans*. Crop Sci. 13: 528-530.

Chopra, V.L. 2000. *Plant Breeding: Theory and Practice*. Oxford & IBH Co. Pvt.   Ltd., New Delhi.

Copp, L.G.L. 1957. Bulk and Pedigree Method of Wheat Breeding; Wheat Information Service No. 5, Biol. Lab., Kyoto Univ., Kyoto, Japan.

Grafius, 1965. Short cuts in plant breeding. Crop Sci. 5: 377.

Hayes, H.K., Immer, F.R. and Smith, D.C. 1955. *Methods of Plant Breeding*. New York: McGraw-Hill Book Co.

Hoffmann, W., Mudra, A. and Plarre, W. 1971. Lehrbuch der Pflanzenzuchtung Landwirtschaftlicher Kulturflanzen, Band 1, Hamburg: Ver. Paul Parey.

Ikehashi, H. 1977. *New procedure for breeding photoperiod sensitive deep water rice with rapid generation advance*. In: Proc. Of Workshop on Deepwater Rice, pp. 45-54. IRRI, Los Banos, Philippines.

Jensen, N.F. 1952. *Intra-varietal Diversification in Oat Breeding*. Agron. Jour. 44: 30-34.

Poehlman, J.M., and Borthakur, D.N. 1969. *Breeding Asian Field Crops*, New Delhi: Oxford & IBH Publ Co. Pvt. Ltd.

Sharma, J.R. *Principles and Practices of Plant Breeding*. Tata-McGraw-Hill Publ. Co. Ltd., New Delhi

Simmonds, N.W. 1979. *Principles of Crop Improvement*. London: Longman.

Tee, T.S. and Qualset, C.O. 1975. *Bulk Populations in Wheat Breeding: Comparison of Single Seed Descent and Random Bulk Methods*. Euphytica. 24: 393-405.

Method of Breeding Cross-Pollinated Crops

As enumerated in Chapter 3, the following methods are used in breeding cross pollinated (naturally cross-pollinated and often cross-pollinated) crops.

1. INTRODUCTION

Past history of cross-pollinated crops reveals many introductions, of released varieties and component lines in a composite or inbreds, from various countries. Many such cases have been cited in Chapters 4 and 14.

2. SELECTION

The following types of selection are practised in cross-pollinated crops.

2.1 Inbred Line Selection

Such selection has not been much utilised in developing varieties in cross-pollinated crops as in self-pollinated crops. By nature, cross-pollinated crops are heterogeneous and heterozygous. Individual plant selection followed by selfing would lead to segregation and cause progeny to deviate from the parental type. The outcome of selection thus would have reduced vigour and productivity. But single plant selection or inbred line selection has been used to develop inbreds. These inbreds are used to develop hybrid or synthetic varieties. The main steps to develop inbreds in cross-pollinated crops by pure line selection are given in Figure 7.1.

2.2 Mass Selection

Other procedures and basis of mass selection in cross-pollinated crops are the same as discussed in Chapter 5. In mass selection a number of plants are selected on the basis of their phenotype, and open pollinated seed from these plants is bulked to grow the next generation. In cross pollinated or often cross-pollinated crops, mass selection has been used to develop varieties and to improve populations extensively. Due to natural cross pollination, there is continuous flow of genes between different individual plants. Since individual plant in such a population is heterozygous, selection of such plants would give an array of variability in the succeeding generations. Therefore cross pollinated populations respond to selection for many cycles of selection. For example, a maize population had 4.70 per cent oil and 10.92 per cent protein. Leng (1961) reported response to selections for high and low types for oil and protein. The populations responded continuously (Table 7.1).

Table 7.1 Result of mass selection in maize population with 4.7 per cent oil and 10.92 per cent protein (data for every 10th cycle)

Generation	High oil	Low oil	High protein	Low protein
Base population	4.70	–	10.92	–
10th	7.38	2.67	14.26	8.65
20th	8.51	2.07	15.66	8.68
30th	10.21	1.44	18.16	6.50
40th	10.16	1.24	22.92	7.99
50th	14.30	1.00	19.50	5.15
60th	14.38	0.77	22.84	4.96

This shows that after 60 cycles of selection for extreme types, a population with 4.7 per cent oil was changed into as high as 14.38 per cent and as low as 0.77 per cent oil. So was the case with protein.

Similar is the history of sugar content in sugarbeet. The progenitors of sugarbeet had about 4 per cent sugar. The sugar content of varieties in 1747

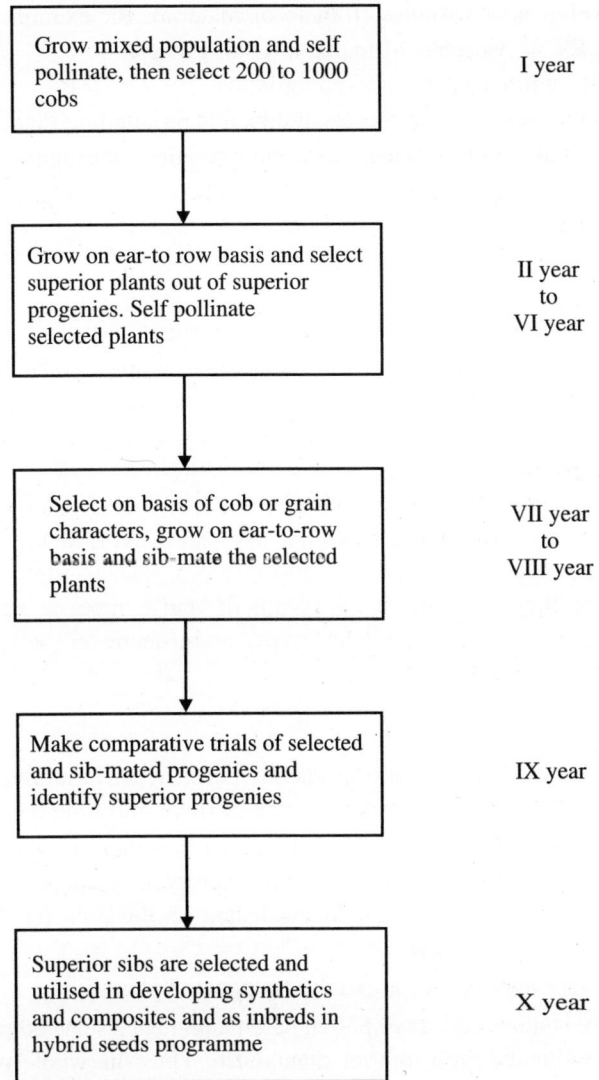

Figure 7.1 Selection scheme for developing inbreds in maize.

was about 7 per cent which rose to 8.8 in 1838, 10.1 in 1868, 16 in 1912 and 18 per cent in modern day varieties. This was a simple act of mass selection, which made this gain possible.

Uses: Mass selection is more effective in cross-pollinated crops due to existing heterozygosity and continuous amalgamation of genes under cross pollination in nature. Mass selection, thus may be practised to:

(i) Develop new varieties from local material, for example, Jaunpuri and KT 41 varieties of maize in Uttar Pradesh.
(ii) Purify unimproved or mixed varieties.
(iii) Maintain a variety where seed production programme is not organised.
(iv) Make minor improvements in existing varieties, for example, uniformity in duration and cob characters in synthetic and composite varieties of maize.

Advantages

1. Mass selection is a simple and easy method as it does not involve crossing, selfing or progeny testing. It is simply selection of plants based on the phenotype (appearance) of the plants.
2. It is a quick method of improving a variety and its simultaneous seed multiplication.
3. It is very efficient in improving the population for the visibly identifiable characters with high heritability, like pant height, maturity duration etc.
4. Minor improvements in the synthetic and composite varieties for uniformity of duration, plant type, grain characters, etc, are done with the greatest ease.

Limitations

1. Mass selection is based on phenotype only, and improvements for characters with low heritability are difficult and time consuming.
2. It is not suitable for self-pollinated crops because homozygous individuals, though selected from heterogeneous population, breed true. Thus the best plants once selected set the limit that no further selection is beneficial.
3. Not very suitable for asexually propagated plants.
4. There is no control over pollen parent and hence selected plants may get pollinated with inferior plants also. Thus the whole purpose of selection is defeated. Also, only the female parent gets considered for selection.
5. Strict selection for uniformity would lead to inbreeding depression.
6. Variety developed by mass selection remains heterogeneous, and hence continuous selection is desirable to maintain it.

In spite of these limitations, mass selection is still being used for breeding new varieties like variety Phule Panchami in sorghum, Phule Dhanwantary in cotton were developed in 2012.

3. RECURRENT SELECTION

In an effort to develop superior inbred lines and to improve population, recurrent selection was developed. In a crude form Hayes and Garber (1919) and East and Jones (1920) gave the ideas. Later Jenkins (1940) detailed the method. Hull (1945) named it as recurrent selection and defined it as reselection generation after generation with interbreeding of selects to provide for genetic recombination.

The method consists of selfing the selected plants and at the same time evaluating them. Superior plants are propagated by selfed seed and are crossed inter se. The resulting inter cross population is again utilised as base population to initiate the second cycle of selection. The cycles of selection and crossing continues till desired results are obtained.

The advantage of recurrent selection is that the ceiling performance is set by the most favourable combination of genes contained in a group of foundation plants and not by a single plant. This system also maintains high genetic variability and allows recombination among different lines.

This method was originally proposed for the improvement of maize inbreds by ways of concentrating favourable genes, but has been used for improving fibre strength of cotton and sugar content in sugarbeet. The method has also been suggested to concentrate certain genes in self-pollinated crops where male sterility exists to aid out crossing.

There are four main types of recurrent selections:

3.1 Simple Recurrent Selection

As the name indicates, it is simple and may be taken as an extension of mass selection. A number of plants are self pollinated and superior ones are selected plants are planted in the second year on plant-to-row basis and all possible crosses are made among lines. Crossing of plants within a line is not done. Crossed seeds of all crosses are composited to make another population where similar cycle is initiated as in the first and second years (Figure 7.2). This cycle of crossing and selection is repeated till the desired results are obtained. If superior plants can be identified before flowering, crossing which is done in the second year can be done in the first year itself. For example, while concentrating genes for resistance against downy mildew or leaf blight in maize, resistant plants are identified in early vegetative stage. These plants may be inter-crossed in the first year itself. Thus it is apparent that in simple recurrent selection, selections of plants are made on the basis of phenotype. Combining ability is not tested. Therefore it is also known as recurrent selection for phenotype.

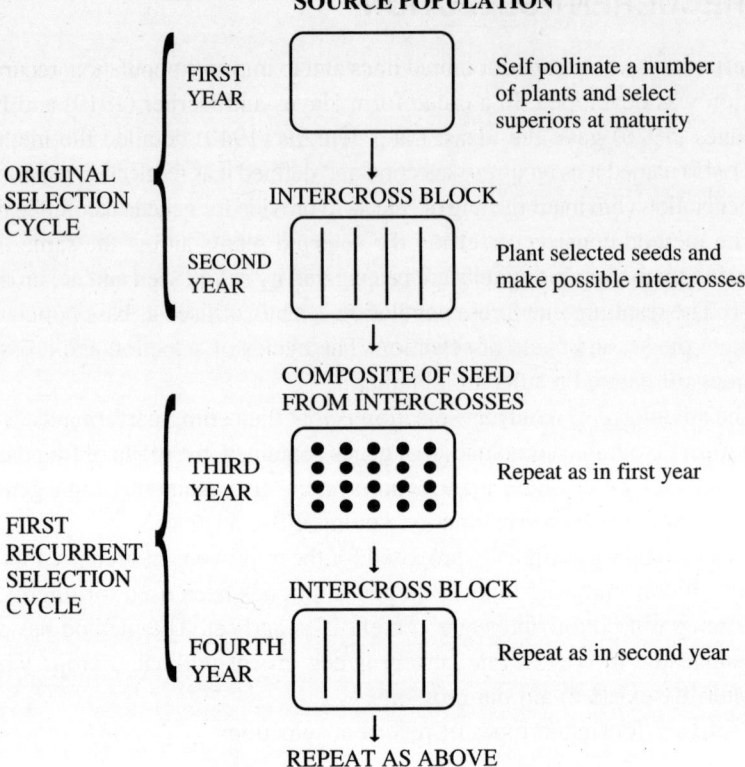

Figure 7.2   Main steps in simple recurrent selection.

3.2 Recurrent Selection for General Combining Ability

In this system, a number of superior plants are selected from the source population. These selected plants, called **So** plants, are selfed and at the same time crossed to a heterozygous tester of broad genetic base. The selfed seed is kept in cold storage. The crossed seed of **So** plants with tester is used to evaluate the combining ability of various So plants. The **So** plants with good performance are grown in the third year from their selfed seed kept in store. They are inter-crossed in all combinations and the composite of inter-crossed seed is they used to establish a new population for further selection. This cycle may be repeated.

3.3 Recurrent Selection for Specific Combining Ability

This method was originally proposed by Hull (1945) to improve specific combining ability. In procedure this method is the same as recurrent selection for general combining ability, except that the tester is inbred line. Because of the narrow base of the tester, the differences in the performance of **So** plants in crosses are due to specific combining ability alone. The limitation of this method has been pointed out as the failure of purpose when the tester turns susceptible or inferior in some respects.

3.4 Reciprocal Recurrent Selection

Comstock, Robinson and Harvey (1949) suggested this method to improve simultaneously two populations for general and specific combining abilities. The foundation material in this method is from two diverse sources, A and B. one population serves as tester for the other. The procedure may be out lined as under (Figure 7.3).

1st year: Outcross about 200 source A plants with a few (say 5) B plants at random. Similarly 200 source B plants with a few A plants should be outcrossed. Self all plants of A and B used as pollen parent in these crosses.

2nd year: Test the crosses made with A and B pollen parents separately.

3rd year: Plant the selfed seed of those A and B plants whose progenies were superior in the 2nd year's trial.

4th year: Year and onwards: Repeat procedure of 1st, 2nd and 3rd years till the desired result is obtained.

4. HYBRIDISATION

The main aim of hybridisation in cross-pollinated crops is to look for heterotic crosses combination i.e. hybrids. Similar to self-pollinated crops, hybridisation in cross-pollinated crops involves crossing two or more inbreds.

4.1 Single Cross

Generally the maximum amount of heterosis is obtained in single cross. But due to weak inbreds, little amount of seed is produced. This is why single crosses have not much been used as commercial variety in the past. The total number of single crosses produced by a given number of inbreds (n) can be calculated by the formula $n(n-1)$ and excluding reciprocals by $\dfrac{n(n-1)}{2}$.

Thus with four inbreds, 12 single and 6 straight crosses (without reciprocals) can be made.

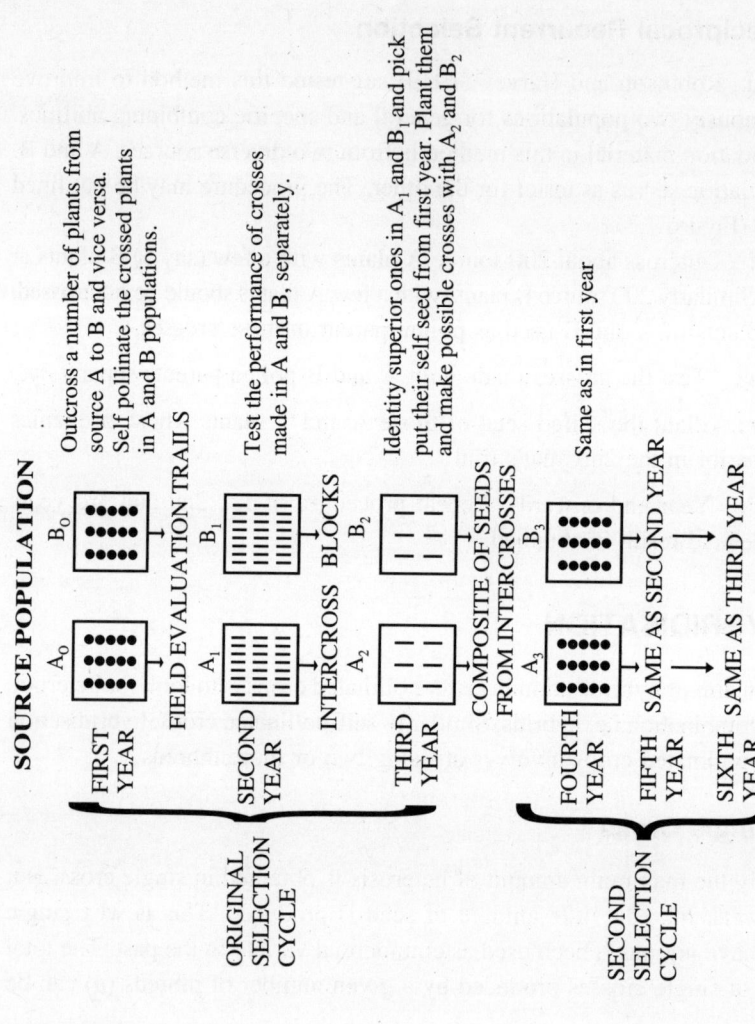

Figure 7.3 Main steps in reciprocal recurrent selection.

4.2 Double Cross

This is the cross between two single crosses and involves four different inbreds. Thus if A, B, C, D, represent inbred lines, one of the possible single crosses can be represented by A x B and one of the possible double crosses by (A × B) × (C × D). Two single hybrids are crossed together to produce a double cross hybrid. The number of possible double crosses can be calculated by the formula:

$$\frac{n(n-1)\,(n-2)\,(n-3)}{8}$$

If n equals number of inbreds, with 4 inbreds, 3 double crosses are possible.

$$\frac{4(4-1)\,(4-2)\,(4-3)}{8} = \frac{4 \times 3 \times 2 \times 1}{8} = 3 \text{ double crosses}$$

4.3 Top Cross

This is also known as inbred-variety cross (Variety X Inbred). Thus top cross is the cross of an inbred with an open pollinated variety. Also as the cross of a single cross with another inbred, (A × B) × C is known as double top cross or three-way cross. This is used not only for developing a hybrid but to test the combining ability of the inbred. Usually the variety or single cross, which is vigorous, is used as female and inbred as male to maximise the seed yield.

    Use: (1) Testing of combining ability, (2) As double top cross hybrid variety.

4.4 Synthetic Cross

This is the cross of a number (4 – 10) pre-tested inbreds and is done by open pollination in isolation. Seeds of these inbreds are mixed in equal proportion and sown in isolated plots. Natural cross pollination takes place and the harvested produce becomes synthetic cross. Synthetic cross is used in testing combining ability and for developing synthetic varieties in crops where pollination is difficult.

4.5 Poly Cross

The term poly-cross was proposed by Tysdal et al. (1942) for the progeny from the seed of a line that was subject to out crossing with other selected lines growing in the same nursery. This differs from synthetic cross in that the lines involved are not pre-tested.

Use: Poly-cross is usually used in forage crops where crossing is difficult. Selected lines or clones are planted in isolation in a field called poly-cross nursery. Planting is done in such a manner as to encourage random inter pollination. Seeds produced by several replicates of the same clone or line are assembled, tested and released as variety if found promising.

5.   TYPES OF VARIETIES AND THEIR DEVELOPMENT

Cross-pollinated crops have adapted to a system where heterozygosity leads to superior performance. Based on this consideration, only hybrid varieties have been developed in cross-pollinated crops. A hybrid variety is one where heterozygous material is used to raise the crop. Depending on the type of cross, varieties of cross-pollinated crops may be of the following types:

5.1   Single Cross Hybrid

The first attempt to use single varietal cross hybrids as commercial variety was initiated in maize in the late nineteenth century by W.J. Beal at Michigan in U.S.A. But it did not become popular. It was G.H. Shull (1909) who made a systematic approach to produce and use F_1 hybrids of two inbreds. This also did not succeed due to the following reasons:

(i) Available inbreds were poor and the resulting hybrid was also not very superior to the open pollinated variety.

(ii) The F_1 seed was small and unattractive.

(iii) The F_1 seed was costly. This was because the female parent was a poor yielding inbred. In addition 1/3 to 1/2 field was planted with the male parent and thus per hectare yield of crossed seed was very less, making it expensive.

But there are a number of other cross pollinated and often cross-pollinated crops like castor (GCH3), onion, cucurbits, carrot, cotton (H4, DH-7), pearlmillet or bajra ((HB1, Saburi), pigeonpea (ICPH-8), rapeseed (PGSH1), sorghum (CSH1), sunflower (BSH1), maize (Co6, HM12), mango, etc, where single crosses are being used (Table 9.3; 14.4; 14.5). Single crosses are also being used in some self pollinated crops taking help of cytoplasmic male sterility like rice (PRH10, TNAU Rice Hybrid Co 4, CRHR 32, JRH 8, Indira Sona, NSD-2).

5.2   Double Cross Hybrid

D.F. Jones (1918) suggested this to make hybrid maize economically feasible. In a double cross hybrid the seed used for commercial planting is produced on a single cross seed parents (Figure 7.4) that yield two to three time as much as any inbred line. Pollen is produced in abundance by the other single cross used as pollinator, and so less land has to be given to it than if the pollinator was an inbred. In addition, the seed is normal in size and shape and produce vigorous seedlings. Thus these are superior to single cross hybrids.

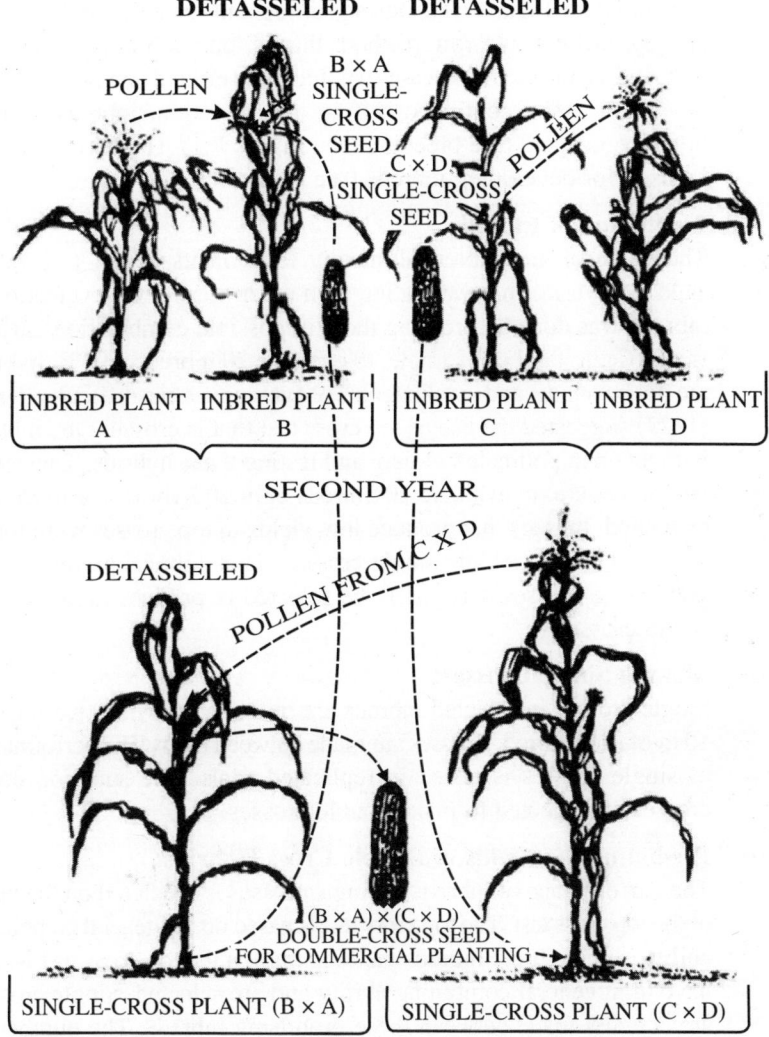

Figure 7.4 Procedure for production of double cross hybrids in maize.

The first commercial double cross was produced in 1921. Now double cross hybrids are a commercial success. The following steps are involved in the development of double cross hybrids.

(i) **Development of Inbred Lines:**

Inbred lines are developed from the genetically broad based populations like open pollinated varieties, composites, synthetics and germplasms by self pollination for five to seven generations, that is, pure line selection (see Figure 7.1). During selfing and selection process these are grown by ear-to-row method, that is, one cob is grown in one line, and vigorous and disease free lines are selected and further selfed or sibbed. After continued selfing and selection, inbreds become homozygous and true breeding (see Figure 7.1). Haploids may also be used for developing inbreds (see Chapter 11).

(ii) **Evaluation of Inbreds:**

The value of any inbred ultimately rests on its ability to produce superior hybrids in combination with other. Earlier the evaluation of inbreds was done by crossing them in possible combinations, which is a difficult job. For example, to evaluate 20 inbreds, 190 F_1 hybrids are to be made and evaluated. To overcome this difficulty Davis (1927) suggested the use of top cross test that is crossing the inbreds with an open pollinated variety and testing these hybrids. Thus, with this procedure, to evaluate 20 inbreds only 20 hybrids are made and evaluated. Inbreds that produce low yields in top crosses were found to produce low yielding single crosses. Thus inbreds producing top yielding to crosses may safely be selected to produce high yielding single crosses.

(iii) **Making Single Crosses:**

Single crosses in selected inbreds are made by growing two lines of 10 m of each inbred. Crosses are made between them. The performance of single crosses is tested in replicated trials. The superior single crosses are selected to make double crosses.

(iv) **Predicting the Yields of Double Cross Hybrids:**

The performance of inbreds in single crosses is predicted on the basis of the top cross test. The top cross test is based on the general combining ability of the inbred. Inbreds yielding high in the top cross test would have high general combining ability and are selected. Single crosses are actually made between these promising inbreds. The number of double crosses, which can be made using the single crosses, is large. This is again a difficult job, and the number of actual crosses is reduced on the basis of prediction. Jenkins (1934) gave as a basis for prediction the average of non parental single crosses. For example, with A, B, C, D, inbreds A × B, A × C, A × D, B × C, B × D and C × D single crosses are possible. The predicted yield of a double cross (A × B) × (C × D) would be the average of all single crosses leaving A x B and C x D, that is, parental.

$$\text{Predicted yield} = \frac{(A \times C) + (A \times D) + (B \times C) + (B \times D)}{4}$$

(v) **Testing Experimental Hybrids:-**

After the double cross yields have been predicted, a number of highest yielding double crosses are chosen and actually made. These hybrids are tested in the field in many locations for a few years. The superior ones are released for cultivation. The seed size of double cross maize is bigger than that of single cross and the amount of hybrid seed produced is also more and comes cheaper. Due to these two reasons double cross hybrids are used commercially.

Many double cross hybrids have been released in maize for cultivation, for example, Ganga 1, Ranjit, Deccan, VL 54, Ganga 3, Himalayan 123 (see Table 14.2 of Chapter 14).

5.3   Double Top Cross Hybrid

Double top crosses are the hybrids between a single cross and an open pollinated variety. The major reasons for the utilisation of the double top crosses are the perfect utilisation of the hybrid vigour of single cross and the quality and adaptability of open pollinated variety. Some double top crosses have been released as varieties, for example, Ganga Safed 2, Hi-Starch, Ganga 4 and Ganga 5 (Table 14.2).

5.4   Synthetic Varieties

This is a variety that is maintained by open pollination following its synthesis by hybridisation among a number of tested genotypes. Only genotypes that combine well with each other in all combinations are put to form the synthetic variety.

(i) **Purpose:**

If we use the F_2 seed of a hybrid and a synthetic, it has been observed that the inbreeding depression is more in hybrids. This indicates that hybrids have more of non-additive (overdominance and epistasis) gene action and synthetics have more of additive gene action. Wherever hybrid varieties are costly or pollination is difficult, synthetic varieties are used, for example, in forage grasses.

Synthetic varieties yield more than inbreds but not more than single or double cross hybrids. But they have the advantage of being cheaper and of not showing any reduction in yield for three to four generations. Thus farmers can use same seed for 3-4 years whereas in single or double cross hybrids every year fresh seed has to be purchased. Synthetic varieties are better than open pollinated varieties as they show at least some amount of heterosis and also are more adapted to changeable growing conditions.

Hayes and Garber (1919) were apparently the first to suggest the possibility of the commercial utilisation of synthetic varieties.

(ii) **Principle behind Synthetic Varieties:**
Since the best combining inbreds are put to form synthetics, heterosis is utilised in the first year to the fullest extent. In subsequent years the decline in yield is less in random pollination. The decline may be calculated by Wright's (1922) formula:

$$F_2 = \frac{F_1 - (F_1 - P)}{n}$$

where F_2 is the expected performance of synthetic variety in the next generation and F_1 the mean performance of F1s from all combinations of inbreds and P the mean performance of n inbreds. Thus in each subsequent generation yield reduction would be 1/2 of heterosis expressed in F_1.

An important consideration in synthetic cross is the number of inbreds to be used so as to give maximum amount of heterosis. Kinman and Sprague (1945) concluded that the expected yield of F_2 increased with increasing numbers of inbreds lines up to 6 and decreased as the inbreds increased from 6 to 10. If all inbreds were to combine, the optimum inbreds would be closer to 10.

(iii) **Procedure of Development:**

(a) **From So or S1 plants:**   This method is based on the fact that the combining ability of the lines is fixed in the early generation of inbreeding and could be identified by the top cross performance. Thus inbred lines in So or S1 are top crossed to rest their combining ability. The best lines are inter-crossed to produce a synthetic.

(b) **From polycross:**   In some of the small flowered and cross pollinated crops where elaborate hand crossing is difficult, in order to test the combining ability through top crossing, polycross nursery is used.

Poly-crossed progenies are the progenies of a population obtained as a consequence of outcrossing with a selected group of lines growing in the same nursery. The lines are grown at random in a polycross nursery, with a minimum of 10 replications. Free outcrossing takes place and the seed set on a line in all replicates is pooled and the performance is tested (Figure 7.5.).

(c) **From established inbreds:**   Fixed inbreds (4-10) may be top crossed to test their combining ability. The best combiners may be inter-crossed, tested and released as synthetics.

For making synthetics in any of above three systems, an equal quantity of F1 seeds of crosses among the best lines is mixed to make the first generation of the synthetic variety. This is called Synthetic 1 generation. Random pollination of Synthetic 1 produces Synthetic 2 and so on.

(iv) **Synthetic Varieties in Various Crops:**

Synthetics have mostly been used in forage grasses, clover, sunflower and sugarbeet. Till 1965, out of 21 varieties of alfalfa released in USA. 17 were synthetic varieties. Synthetic III variety of cauliflower performs well in India.

6. COMPOSITE

Composites are the advanced generation seed mixture of an inter-varietal or inter racial cross. Compared to hybrid varieties whose fresh seed is to be purchased by farmers every year, in composites the same seed is used for 3-4 years. Thus farmers growing composites purchase the seed one year and used its produce again as seed.

Procedure of development: The basic concept of composite varieties lies in high heterosis in some inter-varietal crosses where inbreeding depression is negligible as explained in the case of synthetics. The steps followed to develop composite varieties are depicted in Figure 7.6.

Some release composite varieties in maize are Amber, Vijay, Jawahar, Kisan, Sona, Vikram, Protina, Rattan, Shakti, Tarun and Vikas. Some released composite varieties in pearl millet are Parbhani Sampada, Mandor Bajra (see Table 14.4). Sunflower has prominent composite variety DSRF-113. Differences and similarities among hybrid, synthetic and composite varieties have been given in Table 7.2.

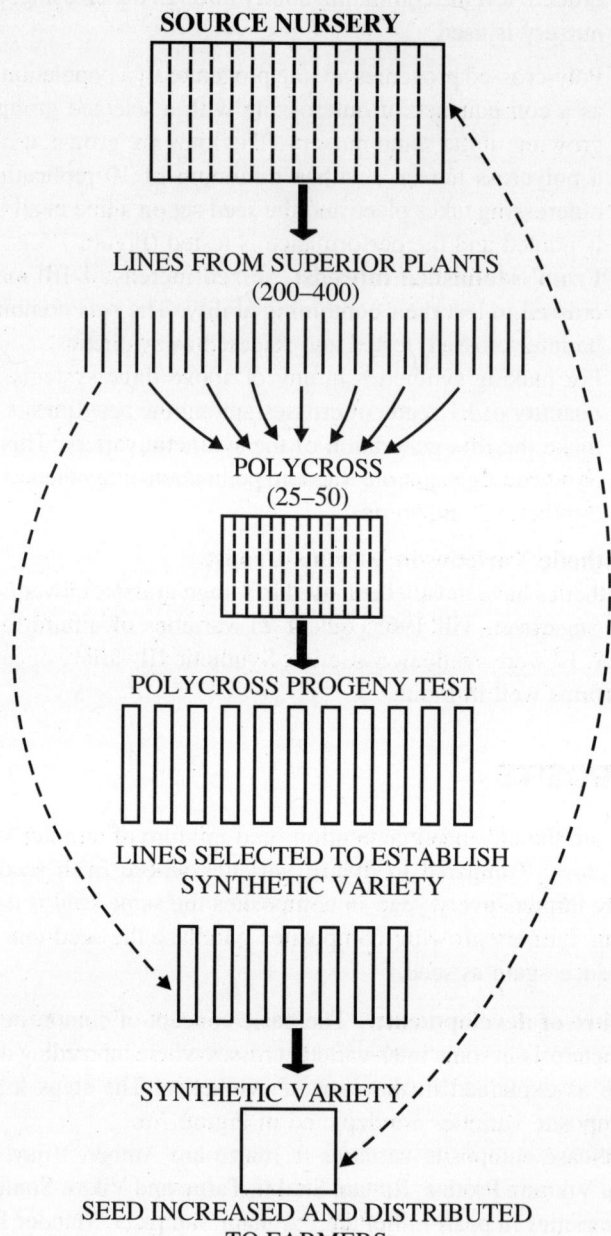

Figure 7.5 Polycross nursery to develop synthetic variety.

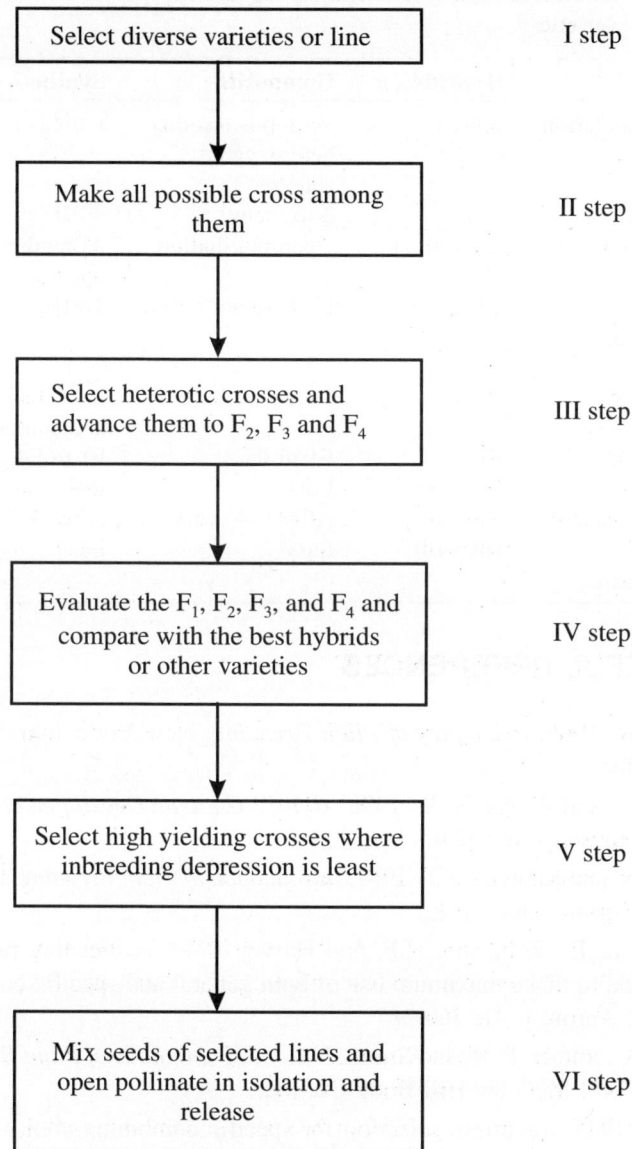

Figure 7.6 Procedure for developing composite varieties.

Table 7.2 Differences and similarities among hybrid, synthetic and composite varieties

Basis	Hybrids	Composites	Synthetics
1. Base population	Inbreds	Varieties or other heterozygous source	Inbreds or clones
2. Parents	2–4	2 to many	4–10
3. Pollination	Controlled	Open pollination	At random, open pollination
4. General combining ability	Tested	Usually not tested	Tested
5. Heterosis	More	Less than hybrid	Less than hybrid and composite
6. Seed used	F1	F1 to F4	F1 to F4
7. Seed cost	More	Less	less than both
8. Seed replacement	Annual	After 3-4 years	After 3-4 years
9. Varietal maintenance	Difficult	Easy	Easy

8. USEFUL REFERENCES

Allard, R.W. 1960. *Principles of Plant Breeding*. New York: John Wiley & Sons, Inc.

Banga, S. S. and Banga, S. K. 1998. *Hybrid Cultivar Development*. Narosa Publ. House, New Delhi.

Briggs, F. N. and Knowles, P.F. 1967. *Introduction to Plant Breeding*. Reinhold Publ. Corpn., New York.

Comstock, R. E., Robinson, H.F. and Harvey 1949. A breeding procedure designed to make maximum use of both general and specific combining ability. Agron. J. 41: 360-367.

Hayes, H.K., Immer, F. R. and Smith, D.C. 1955. *Methods of Plant Breeding*. New York: McGraw-Hill Book Co. Inc.

Hull, F.H. 1945. Recurrent selection for specific combining ability in corn. Jour. Amer. Soc. Agron. 37: 134-145.

Sharma, J. R. *Principles and Practices of Plant Breeding*. Tata-McGraw-Hill Publ. Co. Ltd., New Delhi

Breeding Asexually Propagated Plants

1. VARIOUS GROUPS

There is a group of field, vegetable and fruit crops and ornamental plants which are cultivated using asexual or vegetative parts such as stems, roots, modified flowers, etc. This is a heterogeneous group of plants and can be further divided into four groups for ease in understanding the breeding methods applied to them.

1.1 Group I: Flowerless or Steriles

This group either does not bear flowers or, if at all, these are sterile and do not set seed. For example, many yams, garlic, betel, etc.

1.2 Group II: Apomicts

Apomixis may be defined as the substitution for sexual reproduction of an asexual process, which does not involve any nuclear fusion, as described in Chapter 3. Apomixis has been reported in over 300 genera, yet stable apomixis has been reported only in a few cultivated plants like meadow grass, citrus, mango, raspberry, etc.

1.3   Group III: Flowers but Seed-Set Rare

In this group flowering takes place under specific situations but seed set is limited. For example, potatoes flower only under the short day condition of North India. Sugar cane flowers but sets viable seed only in South India. Therefore the seed is not used for growing crop; rather, stem cuttings are used.

1.4   Group IV: Normal Seed-Set but Propagated Vegetatively

This group consists of fruit trees, ornamentals, etc, which flower and fruit normally but due to long juvenile period, heterozygosity and desirable tree size, propagation is normally done through vegetative means. For example, mango, litchi, citrus, peaches, pear, apple, tea, and most ornamental plants.

2.   BREEDING METHODS

From the above description it is clear that the so called asexually propagated plants have varied nature of propagation. This makes varietal improvement interesting and challenging. The reproductive stages can be utilised for creating more variability through selfing and crossing, and thereafter the vegetative mode of propagation can be utilised in fixing heterosis and in true to type multiplication of heterozygous genotypes clonally.

Some breeding methods are specific to a group but introduction and clonal selection are common to all.

2.1   Group I: Flowerless or Steriles

Only two methods of breeding are used in this group of plants. These are (1) Clonal selection, and (2) Mutation breeding.

2.1.1   Clone and Clonal Selection

In vegetatively propagated plants, plant population may be traced back to a particular vegetative progeny of a particular single plant. A clone may be defined as a group of plants derived from a single plant by vegetative propagation. In other words, all the vegetative progenies of a single plant make a clone. For example, all trees of Dussehari mango available today can be traced back to a single plant of that variety in Dussehari village near Lucknow. Similarly, all trees of Nagpur Seedless Orange can be traced back to Tree No. 182 in Nagpur.

2.1.2 *Characters of clone*

(i) Clonal population is homogeneous. Since each individual plant of a clone is a mitotic derivative of the same plant, they all look similar. Because this is the same tissue of the original plant, which has multiplied into a group of plants, each plant has the same genotype. Any difference in the phenotype of two plants of the same clone would be only environmental as in the case of pure lines.

(ii) Individuals of a clone are heterozygous. All the individual plants of a clone are alike but each individual is heterozygous. Genetically a clone is heterozygous and homogeneous, that is, each plant is heterozygous but the population is homogeneous. Since the parental plant is heterozygous, it would give the same heterozygous plant in vegetative propagation. If propagated through seed it would not breed true but would segregate like a hybrid plant. For example, all plants of grafted Dussehari variety of mango look alike and bear similar type of fruits. But plants grown from Dussehari seed would be different. This shows that Dussehari is not homozygous but heterozygous, and would not breed true. Unlike a vegetatively propagated or grafted plant, a seedling plant comes out of the seed where phenomenon of genetic recombination and segregation operate. That is why each seedling plant is supposed to be different from its mother plant.

(iii) Clones are stable. Like any pure line variety, the clones retain the original characters of the variety. Even after many years of cultivation, sugar cane, potato and other such crop varieties remain stable.

(iv) Mutation is the only means of creating variability. Somatic mutation is the only means through which genetic variability is created. The commonest expressions of such variability are bud-sports and chimera. If mutation occurs in cells from which a bud develops, it would form a bud-sport and on isolation and multiplication would give a new strain or variety. Many varieties have been developed using bud-sports. For example, the Russet Burbank variety of potato is a bud-sport from Smooth Burbank.

(v) Clone is propagated vegetatively. A clone is always propagated and maintained vegetatively. This is the major point where a clone differs from the pure line of self-pollinated crops and inbred of cross-pollinated crops. Other differences among them are given in Table 8.1.

Table 8.1 Differences among pure line, inbred and clone

Pure line	Inbred	Clone
1. Generated in self -pollinated crops only.	In cross-pollinated crops only.	In asexually propagated crops only.
2. Utilised as improved variety and parents for hybridisation.	Used to make hybrids, and in developing composites.	As variety and in hybridisation also if flowers.
3. These are progenies of a single self fertilised individual.	Progenies of selfed or sibbed heterozygous individuals.	Progenies of a single vegetatively propagated usually heterozygous individual.
4. Produced by natural selfing	Produced either by selfing or sibbing.	Produce by vegetative propagation.
5. Pure line is homogeneous consisting of homozygous plants.	Inbred is homogeneous consisting of more or less homozygous plants.	Clone is homogeneous but consisting of heterozygous plants.

2.1.3 *Clonal Selection*

Clonal selection is the selection of desirable clones from the mixed population of vegetatively propagated plants. The technique lies in selecting and propagating the best clone based on its performance (Figure 8.1.). The vegetative part, that is, the unit of selection, differs from crop to crop as listed in Table 8.2.

A generalised clonal selection involves (1) the collection of possible clonal variability, and (2) critical evaluation of each clone and each member of a clone for yield and quality following artificial inoculation of diseases and pests. The disease free and high yielding clones may be selected, further evaluated and multiplied as a variety. Usually selections within a clone are not effective as between different clones.

Advantages

(i) **Easy maintenance:** Selection developed through clonal method are easily maintained as there is no danger of outcrossing and loss of seed viability, etc. Variation occurring through somatic mutation is the only source of variation and can be taken care of by simple rouging.

(ii) **Very quick:** Contrary to hybridisation and selection methods, it is very quick for developing a variety. A single clone, once identified as promising even in the first year, can be multiplied straight way to

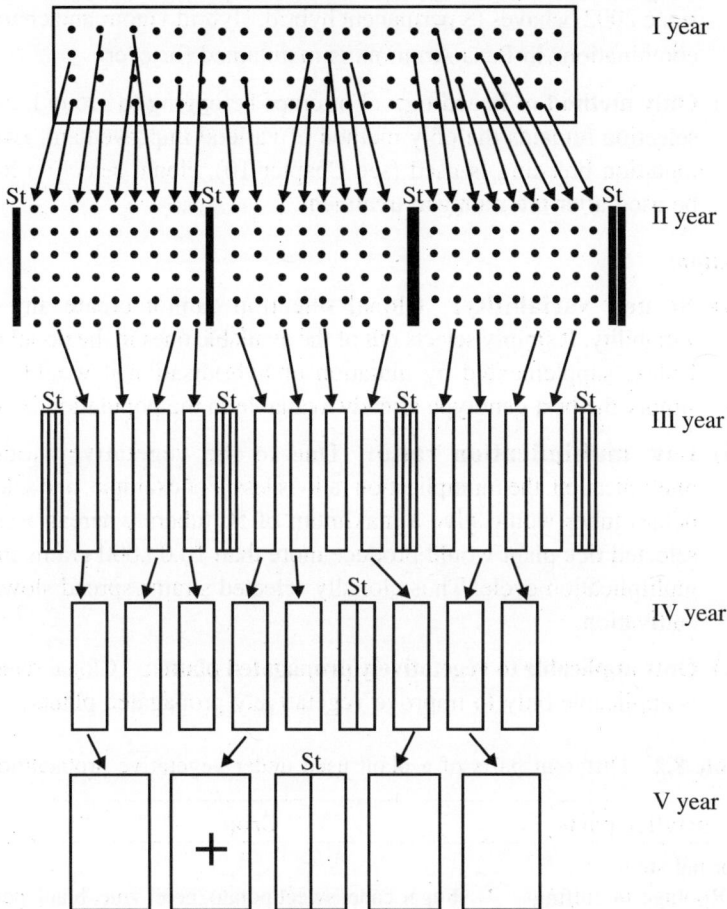

I year

II year

III year

IV year

V year

Figure 8.1 Procedure of clonal selection in asexually propagated plants. After 5th year the selected superior clone marked "+" is selected and multiplied. St is the Standard check.

give a variety. There is no need to purify and fix the character as even the most heterozygous clone would multiply true to type in vegetative propagation.

(iii) **Permanent hybrid:** Heterotic clones on selection can be used as permanent hybrid. Thus heterosis can be exploited for any length of time without the need to produce hybrid seed year after year as in seed crops. For example mango hybrid Pusa Arunima developed by crossing Amrapali × Sensation, crossed once and multiplied by grafting since 2002 behaves as permanent hybrid. Hybrid vigour and character combinations in Pusa Arunima is maintained for ever.

(iv) **Only method of breeding:** For crops belonging to group I, clonal selection remains the only method of varietal improvement. Even if mutation breeding is used (see Chapter 10), clonal selection has to be used after the mutagen treatment.

Limitations

(i) **No new variability:** Clonal selection cannot create any new variability. It simply selects out of the available ones in the population. Unless supplemented by mutation or hybridisation it would select simply the best genotype already available in the population in hand.

(ii) **Low multiplication ratio:** Due to the vegetative mode of multiplication, the multiplication ratio is less. For example, one selected potato tuber would give a maximum of 50 tubers whereas a single selected rice plant would produce more than 10,00,000 grains in one multiplication cycle. Thus clonally selected strains spread slowly in cultivation.

(iii) **Only applicable to vegetatively propagated plants:** Clonal selection is applicable only to improve vegetatively propagated plants.

Table 8.2   Different parts of a plant used under vegetative propagation

Vegetative parts	Crop
1. Normal stem:	
a. Rootage in cutting	Sugar cane, sweet potato, betel vine, black pepper, fodder grasses.
b. Graftage and bud	Mango, citrus, apple and many fruit trees, rose and many ornamental plants.
2. Runner	Oxalis
3. Sucker	Banana, Chrysanthemum
4. Stolons	Mint, Mentha, pineapple, strawberry
5. Tuber	Potato
6. Rhizome and corm	Ginger, Canna, turmeric, Colocasia, yam.
7. Bulbs	Garlic, onion, tulip, lilies.
8. Root and root cuttings	Sweet potato, dahlia, Asparagus, mangoginger.

2.1.4 *Mutation Breeding*

See chapter 10 where this is discussed in detail.

2.2 Group II: Apomicts

Since these groups have some degree of sexual propagation, most methods of vegetatively and seed propagated plants can be utilised, such as:

(i) Clonal selection
(ii) Mutation
(iii) Hybridisation and clonal selection.

A typical example of an apomictic plant *Poa* is given in Figure 8.2. Heterosis can also be fixed in apomictic plants, thus avoiding the production of hybrid seeds, which is costly. This can be achieved by making the hybrid apomictic.

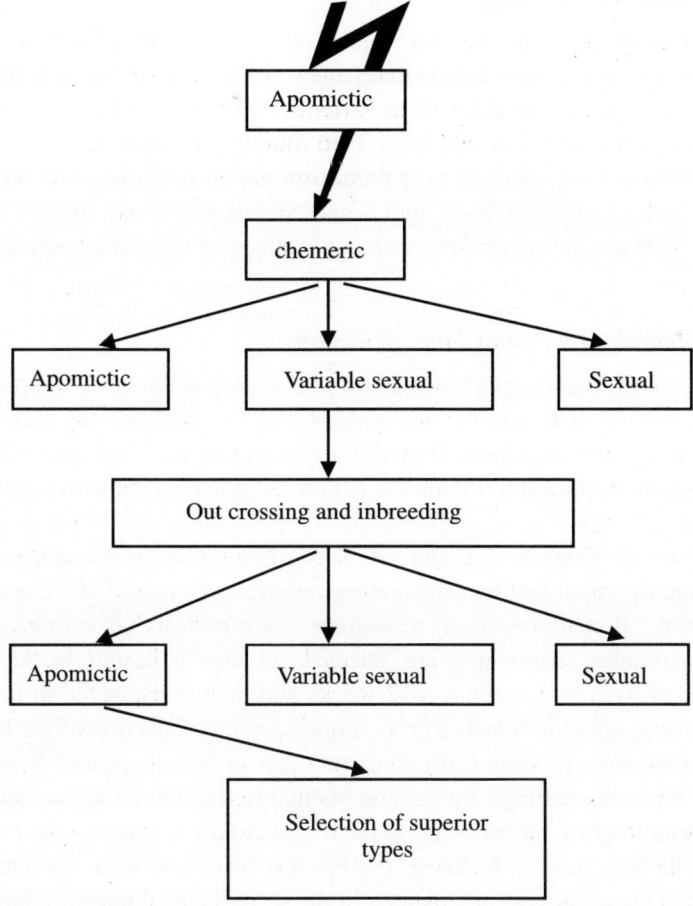

Figure 8.2 Induced transitory sexuality in *Poa partensis* and thereby breeding for better types.

2.3 Group III: Flowers but Seed-Set Rare

The following methods need critical consideration in this group of plants.

2.3.1 *Clonal Selection*

Straight clonal selection has not been of much use in sugar cane but in potato Kufri Red was selected as resistant clone from variety Darjeeling Red Round. Clonal selection has mostly been used to isolate virus the free clones of a variety. But once the existing clonal variability is exhausted, straight clonal selection is not of much use.

2.3.2 *Mutation Breeding*

Mutation breeding in potato and sugar cane is of considerable importance owing to non-flowering of some clones and sterility of other. Somatic mutants obtained may be directly released as varieties. One variety of potato has been developed using mutation breeding. Two mutants of sugar cane 'TS-1' and 'TS-2' (Figure 8.3.), induced by gamma irradiation of variety CO 419, are thicker, high yielders and have high sugar (Shama Rao et al., 1978). CO 997 and CO 6608 are red-rot resistant mutants induced by irradiating with gamma rays.

2.3.3 *Hybridisation and Clonal Selection*

This is now the most important method of breeding sugar cane and potatoes wherein two or more parents are crossed and F_1 seedlings are allowed to grow. Clones (sets and tubers) from the selected plant are advanced for further testing and multiplication. Figure 8.4 depicts a typical example of sugar cane breeding

Sugar cane flowers well and sets seed at Coimbatore and hence at the Sugarcane Research Institute sugar cane varieties are crossed in October and November. Limited crossing is also done in northern Indian stations. After about two months, seeds (fluffs) are collected and sown in the nursery. Seedlings are allowed to grow for about 6-12 weeks and then transplanted in the first field nursery, either in bunches or as single seedling. Selection of seedling is made on the basis of vigour, tillering and other visible characters. Sets from stem of selected seedlings are cut and planted in the first clonal nursery and are allowed to grow for one year. Selection of clones is made on the basis of vigour, tillering, height, thickness, disease resistance and sugar content. Sets of selected clones are cut and planted in the second clonal nursery. Selection is again made and superior clones are thoroughly tested in multi-location trials. The superior-most entries are released for cultivation. The main point

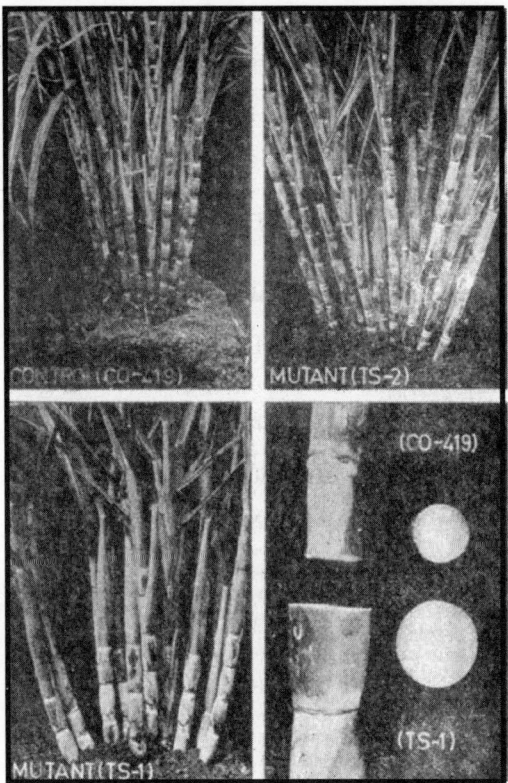

Figure 8.3 Radiation induced sugarcane mutants; TS-1 thick canes, and TS-2 high tillering type (Courtesy Dr. H. K. Shama Rao).

is that selection is practised in F_1 itself. Because sugar cane is highly heterozygous even is varietal crosses, tremendous variability is obtained in F_1 itself.

The same is the case with potato, which is highly heterozygous, and autotetraploid. Crosses are made at the Central Potato Research Institute, Shimla, in summer months. In September-October F_1 seedlings are grown at Jalandhar in boxes and vigorous seedlings are transplanted in pots after a month. By January tuber from selected plants are harvested and stored. These are grown the following winter in the first field nursery and selection is made based on tuber characters and yield. The process is repeated and test locations increased. The superior most selections are released for cultivation. Figure 8.5 depicts an example of potato breeding where, after crossing, clonal selection is practised. Recent varieties of potato such as Kufri Sadabahar, Kufri Pushkar, and Kufri Frysona are the result of such a development programme.

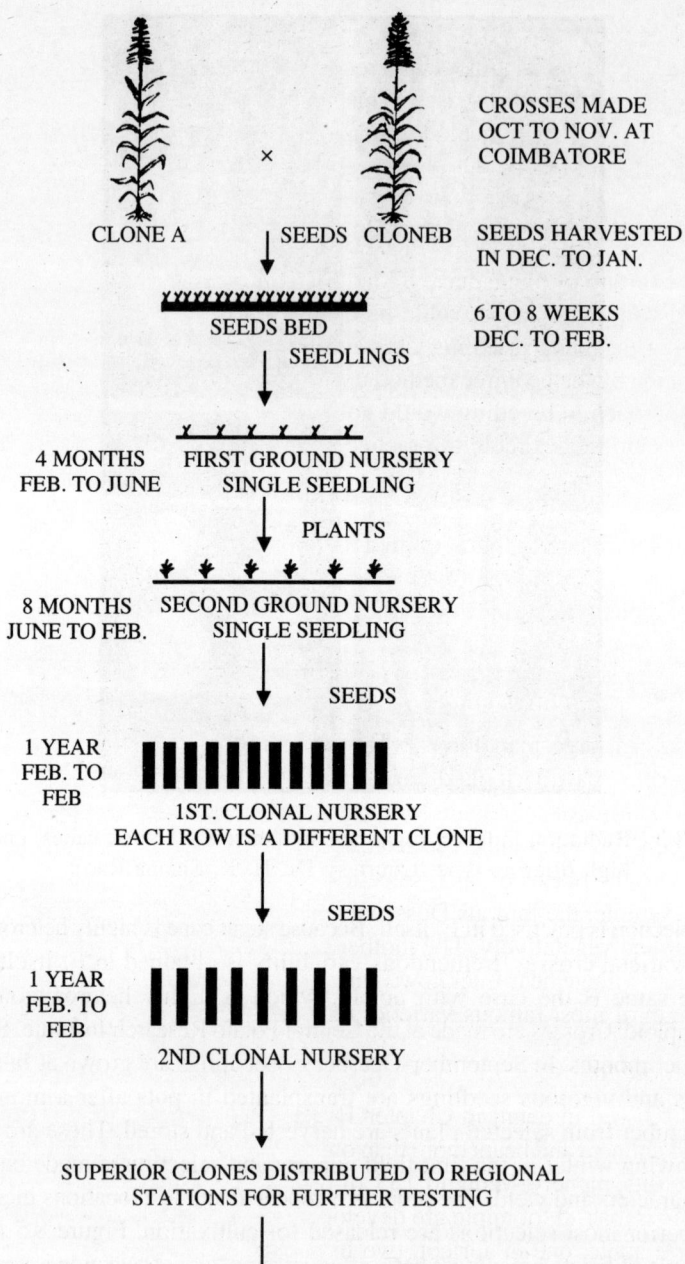

Figure 8.4 Procedure of sugar cane breeding

2.3.4 Haploid Breeding

Potato is tetraploid but its diploid wild relative *Solanum demissum* has resistance against many pests and diseases. This resistance can be transferred using dihaploid of potato. Other methods of haploid breeding are described in Chapter 11.

2.4 Group IV: Normal Seed-Set but Propagated Vegetatively

Since these group of plants have a long juvenile period, even after selfing or crossing a very long period would be needed for advancing the generations. For example, the mango seedling takes at least 5 years to flower, thus advancing the generation by the pedigree method up to F_7 would require 35 years. Clearly, such an approach in breeding would not be feasible. Therefore the following methods would be desirable to practice.

2.4.1 Clonal Selection

Chance variation due to mutation in a clone may be selected. Many varieties have been released through this method. For example, Kufri Red variety of potato is a clonal selection from Darjeeling Red Round and Kufri Dewa from Phulwa.

2.4.2 Mother plant Selection

Many times in large plantations where seedlings plants have been planted, superior types may be looked for, since most plants are heterozygous, in seedlings an array of segregants would be obtained. Larger the population for selection better are the results. Therefore commercial plantations and not breeders' nurseries are the places of such selections, the former being fairly large.

For example, the famous Dussehari mango is a chance seedling selected and multiplied vegetatively. The mother plant of this variety is still surviving in Dussehari village near Lucknow. This plant was grown out of seed. Similar is the origin of most famous varieties of mango, guava and apple.

2.4.3 Mutation

This is discussed in detail in Chapter 10. The Alphonso variety of mango is mericlinal chimera and bears more than one type of fruits. A number of varieties in many fruits, namely 'Mangnif 135' of peach, Stella of cherry, Star Ruby of Citrus, Mc Intosh-8F of apple are developed out of bud-sports. Besides, three varieties in apple, one in apricot, two in cherry, one in grape fruit have also been developed through mutation (Broertjes and van Harten, 1978). Nagpur Seedless Orange is a mutant variety released in 2006 as only one plant (Tree No. 182) was spotted in a *Santara* orchard in Nagpur.

2.4.4 *Hybridisation and Selection*

Like other crop plants, hybridisation and selection has been practised in fruit plants with the great advantage of vegetative propagation. Interspecific or inter-varietal crosses may be made to produce large numbers of F_1 where selection can be practised based on yield and quality character. The best F_1 may be multiplied and released for cultivation. In mango two F_1 hybrids, Probha-shanker and Mahmud Bahar, had been released out of cross Bombay × Kallapady in 1953 at Sabour. Another F_1 hybrid of Amrapali and Mallika were developed by IARI (Figure 8.6) out of a cross of the tasty Dussehari of Lucknow and the regular bearer Neelam of Tamil Nadu. Amrapali combines

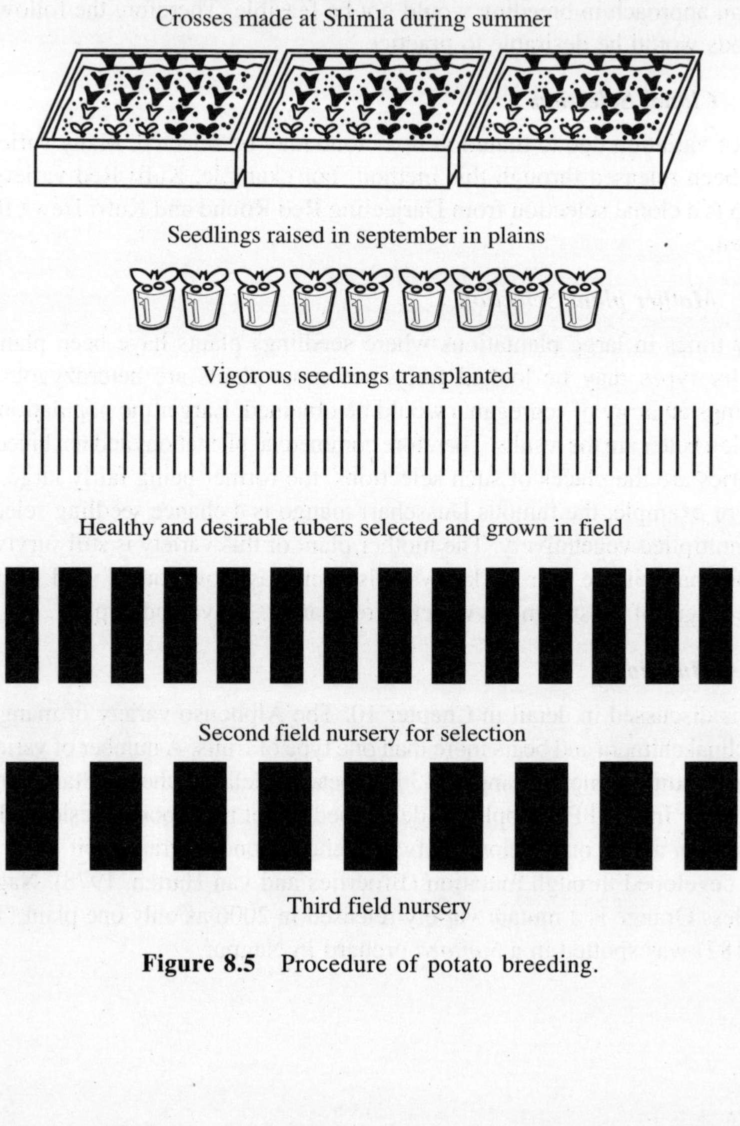

Crosses made at Shimla during summer

Seedlings raised in september in plains

Vigorous seedlings transplanted

Healthy and desirable tubers selected and grown in field

Second field nursery for selection

Third field nursery

Figure 8.5 Procedure of potato breeding.

Figure 8.6   A hybrid variety of mango "Amrapali" (Courtesy: Dr. R. N. Singh).

the parental character and has additional points of reduced plant size, higher pulp and beta-carotene. To reduce the juvenile period, young F_1 seedlings are grafted on older stocks and thus the F_1 comes in bearing soon and selection can be made for fruit characters. Pusa Arunima mango is a cross between Amrapali x Sensation. Rahuri Sugandh is a cross between Totapuri × Kesar mangoes.

2.4.5   *Polyploid Breeding*

This is discussed in detail in Chapter 11. In crops like tea, triploids and tetraploids may be induced and developed as variety. Pierce, a tetraploid variety of grape is very much under cultivation.

Vegetatively propagated plants make a heterogeneous group and their breeding is challenging and interesting. Somehow clonal selection in the beginning has been quite effective in developing better varieties. Of late, hybridisation and mutation breeding have played an important role in developing superior varieties. In a survey in May 1978, by Broertjjes and van Harten (1978), a total of 199 mutant varieties were reported (Table 8.3). Varieties of asexually propagated crops developed in India are listed in Table 8.4.

Table 8.3   Mutant varieties in vegetatively propagated plants elsewhere in the world.

Plant group	No. of commercial mutants in each	Total
1. Root and tubers	Potato- 1*	1
2. Bulbous plants	Dahlia- 23, Lilium-2, Tulip-2, Polyanthes-2	29
3. Pot plants	Begonia-21, Guzmania-1, Streptocarpus-8, Achimenes-8, Azalea-10	48
4. Cut flowers	Alstroemeria-14, Carnation-2 roses-4, Chrysanthemum-78	98
5. Garden plants	Portulaca-7, Bougainvillea-1	8
6. Woody plants	Abelia-1, Malus-1	2
7. Temperate fruits	Apple-4, Apricot-1, Cherry-3, Peach-1	9
8. Small fruits	Black currant-2	2
9. Others	Peppermint-2	2

*Number against each plant indicates the number of mutant varieties.

Table 8.4   In India mutant varieties of asexually propagated crops have been developed mostly in ornamental plants till 2013 (Compiled from: www.mvgs.iaea.org./AboutMutantVarieties.aspx)

Name of crop	Mutant varieties	Name of major varieties
Bougainvillea	10	Jaya, Jayalakshmi Variegata, Los Banos Variegata, Mahara variegata
Chrysanthemum	49	Agnishikha, Alankar, Asha, Batik, Hemanti, Jugnu, Navneet Yellow
Coleus	1	Suphala
Dahlia	10	Bichitra, Black Beauty, Vivekananda
Gladiolus (Gladioli)	2	Shobha, Tambari
Hibiscus (Shoe flower)	2	Anjali, Purnima
Lemon grass (Citronella)	6	Bhanumati, Bibhuti, Saurav, Subir,
Mulberry	1	S-54
Portulaca (Moss rose)	14	Five Petal, Jhumka, Karn Phul, Vibhuti
Rose	14	Abhisankar HT, Angara, Ligh Pink Prize
Sugarcane	5	Co 6608 mutant, Co 85035
Tuberose (Polianthes)	2	Rajat Rekha, Swarna Rekha
Turmeric	2	BSR-1, Co 1
Total	**118**	

3. USEFUL REFERENCES

Barnes, A. C. 1964. *The Sugarcane*. London: Leonard Hill Ltd.

Barua, D.N. 1965. Selection of vegetative clones. Tea Encl. 163. Tocklai.

Broertjes, C. and van Harten, A.M. 1978. Applications of mutation

Breeding Methods. In: The Improvement of Vegetatively Propagated Crops. Amsterdam: Elsevier Sci. Publ. Co.

Pushkarnath, 1961. Potato breeding and genetics in India. Indian J. Genet. 21: 77-86.

Pushkarnath, 1964. Potato in India: Varieties. New Delhi: ICAR.

Rao, J. T. and Narsimhan, R. 1963. Recent currents in sugar cane breeding. Indian Sugar 13(2): 135-137

Shama Rao, H.K. Gujarathi, P.M. and Rannavare, B.M. 1979. Field trials of sugarcane mutants developed by BARC. Proc. 29th Ann. Convention, DSTA.

Singh, L. B. 1960, *The Mango*. Leonard Hill, New York.

Sinha, S. K. 1964. Leaf culture as a tool in potato breeding. Indian Potato Jour. 6: 49-50.

Smith, O. 1977. Potatoes: Production, Storing and Processing. Connecticut: Avi.

Visser, T 1969. *Outlines of Perennial Crop Breeding in the Tropics*. Wageningen.

Heterosis Breeding

1. WHAT IS HETEROSIS

A heterozygous individual resulting from the cross of two unlike parents is a hybrid, which is usually vigorous. This increased vigour is often referred as hybrid vigour or heterosis. Thus heterosis is the phenomenon in which the hybrid of two genetically dissimilar parents shows increased vigour at least over the mid-parental value. Heterosis is just the reverse of inbreeding depression, which results after selfing the plants of the cross-pollinated crops. Kolreuter (1761-66) was the first to report hybrid vigour in the hybrids of tobacco, Datura, etc. Most of the early plant hybridisers including Mendel (1865) had noticed this phenomenon. Dr. G.H. Shull proposed the term heterosis from two Greek word, *hetero* (different) and *osis* (condition), to replace his own earlier word '"stimulus of heterozygosis".

2. TYPES OF HETEROSIS

Depending on the nature, origin, adaptability, reproducibility and non-reproducibility, heterosis can be classified into two: (1) Euheterosis (true heterosis), and (2) pseudo-heterosis (Luxuriance).

2.1 Euheterosis

This is the true heterosis, which is inherited. Euheterosis can be further divided into (a) mutational euheterosis, and (b) balanced euheterosis.

2.1.1 *Mutational Euheterosis*

This is the simplest type of euheterosis and results from the sheltering or shadowing of the deleterious, unfavourable, often lethal, recessive mutant genes by their adaptively superior dominant alleles in population of sexually reproducing cross fertilising organism. Most mutations are lethal, recessive and adaptively inferior but mutational heterosis protects them from quick elimination from population.

2.1.2 *Balanced Euheterosis*

This is the true type of heterosis, which arises out of balanced gene combinations with better adaptive value and agricultural usefulness. It is such type of heterosis, which is used in plant breeding programmes for evolving hybrid varieties.

2.2 Pseudo-heterosis

This is basically a phenomenon by which the crossing of the two parental forms brings in an accidental, excessive and unadaptable expression of temporary vigour and vegetative overgrowth (luxuriance). Luxuriance cannot be utilised in breeding hybrid varieties.

3. MEASUREMENT OF HETEROSIS

Heterosis is measured in three parameters as depicted in Figure 9.1.

Relative Heterosis: This expresses heterosis over mid-parents and is calculated by subtracting the average value of parents from F_1. Relative

$$\text{heterosis} = \frac{F_1 - \text{Mid parent}}{\text{Mid parent}} \times 100$$

Heterobeltiosis: The term was used by Bitzer et al. (1968) to describe the improvement of heterozygote over the better parent of the cross.

$$\text{Heterobeltiosis} = \frac{F_1 - \text{Better parent}}{\text{Better parent}} \times 100$$

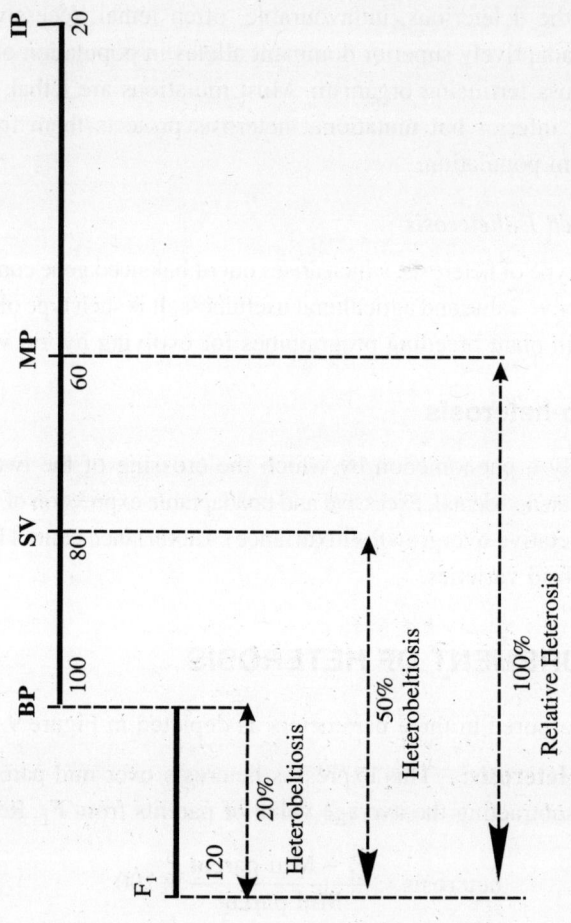

Figure 9.1 Different types of heterosis, where BP= better parent; MP = mid parent value; SV = standard check variety; IP = Inferior parent; F_1 = F_1 hybrid.

Standard Heterosis: This is calculated over the standard check variety to show the superiority of hybrid over the prevalent variety.

$$\text{Standard heterosis} = \frac{F_1 - \text{Check variety}}{\text{Check variety}} \times 100$$

4. BASIS OF HETEROSIS

A. Genetic Basis of Heterosis

Based on the gene action and interaction a number of hypotheses were proposed to explain the heterotic vigour in F_1 hybrid

4.1 Dominance Hypothesis

First proposed by Davenport (1908) and later supported by Bruce (1910), this hypothesis is based on the coverage of the ill effects of recessive genes by the dominant genes in the hybrid. Thus the dominant genes, which are favourable, mask the effect of recessive genes and bring in heterotic effect in the hybrid. For example:

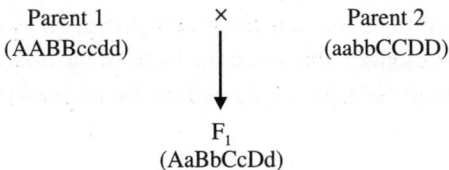

Parent 1 × Parent 2
(AABBccdd)   (aabbCCDD)

F_1
(AaBbCcDd)

Each parent has only two dominant genes but the F_1 has four dominant genes and they mask the effect of recessive genes and show heterosis.

Objections: Two major objections to this hypothesis have been raised.

(a) **Why not true breeding heterotic lines:**   If above hypothesis is correct, it should be possible to recover a plant with genotype AA BB CC DD which is as highly vigorous as F_1 but true breeding.

But the objection was removed by Jones (1917) in his theory 'Dominance of linked genes hypothesis'. He pointed out that a large number of dominant genes distributed over different chromosomes are involved in producing heterosis. The favourable and unfavourable genes may be linked, due to which true breeding plants with all favourable (dominant) loci cannot be obtained.

(b) **Why symmetrical distribution:** If heterosis is due to dominance of independent factors, the F_2 distribution curve for heterotic character should be skewed rather than symmetrical. Because in such a trait the dominant and recessive loci would be distributed according to the expansion of binomial $(3/4+1/4)^n$, where n is the number of genes involved in the expression of the heterotic character.

This objection was removed by Collins (1921) who pointed out that heterosis is governed by a large number of genes, so that even in the absence of linkage skewness would not appear.

4.2 Overdominance Hypothesis

This hypothesis was first given by Shull (1903) and East (1908) independently. The basis of this hypothesis is that the heterozygote (Aa) is superior to both the homozygotes (Aa or aa) and the amount of heterosis depends on the amount of heterozygosity in the hybrid. Thus several names like super dominance by Fisher (1930), interaction of alleles at a single locus by East (1936) and overdominance by Hull (1945) were given. Convincing support to this hypothesis came from Hull (1945) who observed in maize, if two inbreds A and B yielded 10 and 12 quintals, then their F_1 yielded more than 22, say 25 to 27 quintals. This could not possibly be explained with the dominant gene acting even in a completely additive manner, that is, on the basis of the dominance hypothesis. But this could perhaps be better explained on the basis of the overdominance hypothesis.

Objections: Most support to this hypothesis comes from single locus heterosis. Doubts have been raised that what is true for qualitative genes may not hold good for quantitative genes, that is, heterosis in quantitative characters like yield may not be single locus heterosis.

Similarities and Differences between the Two Hypotheses: Both the hypotheses, dominance and overdominance, are based on the assumption of Mendelian interpretation. Both of them agree for inbreeding depression following selfing and heterosis on out crossing. The major point of difference lies in their basic assumption where the first one assumes that heterosis results from sheltering of deleterious effects of the hypothesis it is not sheltering but the complementation between dominant and recessive alleles to produce overdominance. No clear cut proof has come in favour of a particular hypothesis but it is possible that both the systems may be operating simultaneously in the same organism.

B. Physiological Basis of Heterosis:

4.3 Greater Initial Capital Hypothesis

Ashby (1930) studied the inbreds and hybrids of maize and tomato and concluded that in all heterotic crosses, the embryo (initial capital) size of F1 is greater. Now there are many evidences in contradiction of this hypothesis.

4.4 Mitochondrial Complementation

Sarkissian and Srivastva (1967) studied mitochondrial complementation in five inbreds and their single and double crosses of maize. They reported that mixing mitochondria from parents of heterotic crosses only showed complementation, that is, heterotic activity. Since mitochondria are the chemical powerhouses of the cell, complementation at that level makes energy transfer in heterotic hybrids more efficient and is hence expressed as heterotic vigour.

C. Biochemical Basis

Model of Genic Complementation

Robbins (1941) excised root tips from Johannesfeuer and Red Currant varieties of tomato and their hybrids and grew them in solutions supplemented with growth promoting substances-pyridoxin and nicotinamide. It was observed that Johannesfeuer was not able to synthesise pyridoxin, and Red Current nicotinamide and thus responded to their addition. The F_1 hybrid was able to grow without pyridoxin and nicotinamide. These observations suggested that the expression of heterotic advantage is an expression of the activity of the favourable combination of biochemical growth factors produced as the consequence of complementary gene action in the hybrids.

D. Cytoplasmic Basis

Cytoplasm is a limpid fluid rich in RNA and mitochondria which is usually transmitted through the female parent to the offspring. A number of evidences are now available to indicate the interaction of nuclear genes and cytoplasm in the production of phenotype of the individual. Dhawan and Paliwal (1964) crossed Sikkim Primitive (SP_2) and Colorado varieties of maize and reported clear-cut reciprocal differences for heterosis (Table 9.1).

Table 9.1 Yield (kg/ha) of SP_2 and Colorado and their reciprocal hybrids

Pedigree	Yield	Yield % of Colorado	Yield % of SP_2
1. SP_2	615	45	100
2. Colorado	1375	100	224
3. $SP_2 \times$ Col.	63	46	104
4. Col. $\times SP_2$	2972	216	433

Heterosis is a complex expression of hybrids and may be due to the dominance and overdominance phenomena occurring singly or simultaneously. Heterotic advantage is basically the result of efficient and increased metabolic activity of the heterozygote. Interactions of nuclear and cytoplasmic factor also seem involved in the heterotic expression.

5. BREEDING METHODS TO EXPLOIT HETEROSIS

It is clear that heterotic advantage is limited only to the F_1 generation and thereafter it declines. Thus for the full utilisation of heterosis, fresh hybrid seed has to be produced and used every year. The cost of hybrid seed is naturally high. However, hybrid seed could be produced at cheaper rates if certain basic features are available in that crop.

5.1 Basic Features

(i) Availability of superior hybrids to compensate the cost of hybrid seed. Table 9.2. gives the relation between the yield of hybrid and its seed-cost in an advantageous position.

Table 9.2 Percentage of standard heterosis necessary to pay the additional cost of hybrid seed at different commercial yield levels *

Commercial yield (Q/ha)	Additional costs / ha in Rs. **				
	75.0	112.50	150.00	187.50	281.25
6.8	20.0 ***	15.0	40.0	50.0	75.0
13.6	10	15.0	20.0	25.0	37.50
20.4	6.7	10.0	13.0	16.7	25.0
27.2	5.0	7.5	10.0	12.5	18.8
34.0	4.0	6.0	8.0	10.0	15.0
51.0	2.0	4.3	5.7	7.4	10.7
61.2	2.2	3.3	4.4	5.6	8.3
81.6	1.7	2.5	3.3	4.2	6.3

* Modified from Rai (1979)
** Additional cost of hybrid seed over local ones.
*** Standard heterosis required when produce is sold at Rs. 55.1/quintal.

(ii) Easy and efficient pollination control, like monoecious habit, male steriles, incompatibility, etc.

(iii) Strong fertility restorer in case of male sterility.

(iv) Absence of modifier genes in case of self incompatibility so that it does not break in any cross combination.

(v) Complete synchronisation of flowering in male and female parents.

(vi) Free, unrestricted and natural transfer of pollen from male to female parents.

(vii) Good seed setting on the female parent.

(viii) A skilled organised effort for large-scale seed production, certification, processing and well-knitted distribution channel of hybrid seed.

5.2 Types of Hybrids Used

In crop plants, both full and partial utilisation of heterosis has been done using different types of hybrids (Table 9.3). Single crosses give the maximum degree of heterosis and produce the most uniform crop. But due to less amount and small size of seed, its cost becomes high. To overcome these problems, double cross hybrids are used commercially. But with more productive inbreds and higher amount of heterosis, mostly single crosses are being used in maize now.

5.3 Techniques of Producing Hybrid Seeds

The technique lies in producing any category of hybrid seed by efficient pollen control outlined below.

(i) **Hand Emasculation and Pollination**

In hermaphrodite flowers it becomes imperative to use hand emasculation and pollination, though laborious and costly. In Japan 90 per cent of tomatoes are hybrids and the entire seed is produced by hand emasculation and pollination. In India also it seems economic (Khan et al., 1971). When each pollination is capable of producing many seeds as in tomato, brinjal, lady's finger, etc, hand pollination is not costly. In India every seed of Varalaxmi hybrid cotton is produced by hand emasculation and pollination. Variety Lakshmi (*G. hirsutum*) is used as female parent and SB. 289E (*G. barbadense*) as male. In 1977-78, 2176-4 quintals of F_1 seed was produced by hand pollination. Moreover, by skilled workers and improving the method of emasculation and pollination and chemical emasculation methods in rice, wheat and sorghum, hybrid seed can be produced still cheaper.

Table 9.3 Commercial hybrids used in some crop plants. (A to F are parents and OP is open pollinated variety)

Hybrids	Symbol	Crops
A. Full exploitation:		
1. Single crosses	A × B	Bajra, brinjal, carrot, castor, chillies, cotton, cucurbits, onion, sorghum, rice, tobacco, tomato, wheat, maize, sugarbeet
2. Double crosses	(A × B) × (C × D)	Maize
3. three way crosses	(A × B) × C	Sweet corn
4. Double top crosses	(A × B) × OP	Maize
5. Triple crosses	(A × B) × C × (D × E) × F	Cabbage, kale
6. All types of hybrids	–	Potato, sugar cane
B. Partial exploitation:		
1. Synthetic	–	Maize
2. Composite	–	Maize, rape seed
3. Poly cross	–	Grasses

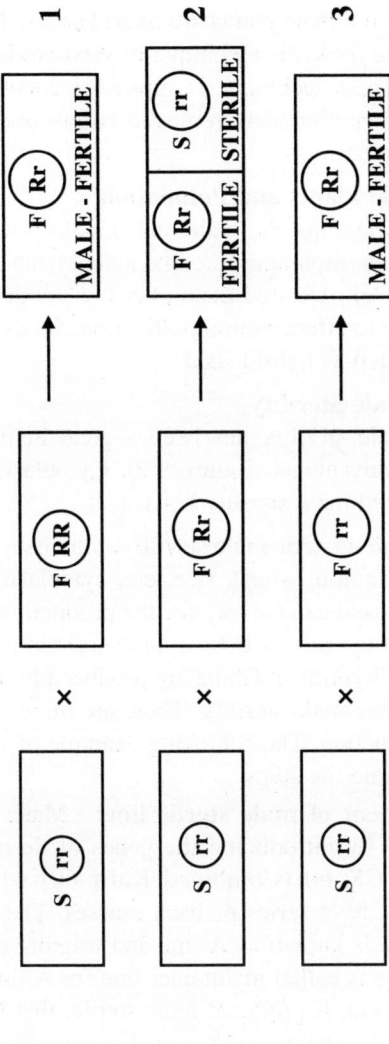

Figure 9.2 Three methods of producing hybrid seed for commercial crop production using cytoplasmic-genetic male sterility. "S" denotes sterile cytoplasmic factor and "R" the restorer gene. All 3 methods are suitable for crops where the commercial product is seed or fruit. Only method 1 is suitable if the commercial product is a vegetative part.

(ii) **Detasseling and Pollination**

Detasseling is the removal of immature tassel for preparation of female (seed) parent in hybrid seed production. For hybrid seed production in maize six rows of female parents are grown after every two rows of male. Detasseling of the female is done and, through natural cross pollination, pollen from the male rows pollinates the female rows. Seeds produced on female plants are hybrid seeds. The entire hybrid maize programme uses this technique of seed production.

An almost similar technique is applied in cucurbits also but the male flowers are continuously removed before opening, to make a plant a female parent.

(iii) **Removal of Male Plants and Pollination**

In dioecious species, that is, male and female flowers on separate plants, like spinach, asparagus, etc, the male plants are removed and only females are left. Desired male plants of the desired variety are grown by the side to effect natural pollination. Seeds borne on female plants are harvested as hybrid seed.

(iv) **Utilisation of Male Sterility**

Utilisation of male sterility has been a great help in hybrid seed production in many plants (Figure 9.2). Cytoplasmic, genetic and chemically induced male sterility is used.

(a) **Use of cytoplasmic-genetic male sterility:**   In many crops like bajra, sorghum, maize, onion, wheat, rice, etc, cytoplasmic-genetic male sterility is being used extensively for the production of hybrid seed. Almost all maize hybrids in U.S.A. and sorghum and bajra hybrids in India, and rice hybrids in China are produced by the utilisation of cytoplasmic-genetic male sterility. There are three steps involved in hybrid seed production. The following example of sorghum (Figure 3.8) would illustrate the steps.

(i) **Development of male sterile line:**   Male sterile lines are developed by introducing the genes of desired male parent into line CK 60 (Combined Kafir 60) which has sterile cytoplasm, by a series of back crosses. The converted male sterile line is known as A line and original counterpart as B line. B line is called maintainer line for A line because cross of A line with B produces male sterile, that is, A line.

(ii) **Maintenance of A line:**   For the maintenance and seed increase of A line, two rows of A are grown alternate to B line in isolation. Seed borne on A lines are collected as seed for A line.

(iii) **Production of hybrid seed:** Single crosses are made between A line and another variety or line (R line) which is reported to have high heterosis and fertility restorer gene (MSC_1). For hybrid seed production A and R lines are either planted in 6:2 ratio or six rows of A after two rows of R. The planting data of these lines is adjusted to synchronise their flowering time. Seeds set on a line are collected and used as hybrid seed, which are fully fertile.

(b) **Use of genetic male sterility:** Genetic male sterility has been reported in crops like barley, carrot, cotton, maize, sorghum, wheat, etc but this does not eliminate manual labour completely as in the case of cytoplasmic male sterility. Since most male sterility genes are recessive (ms ms), on crossing with the maintainer (Ms ms), it segregates into 1:1 ratio of male fertile and sterile. Thus is used as A line, 50 per cent plants must be rouged before flowering, and these would be male fertile.

(c) **Chemical suppression of male flower:** Wittwer and Hillyer (1954) observed that spraying dilute solutions of maleic hydrazide or ß - NAA (100 ppm) on young seedlings of *Cucumis pepo* suppressed the formation of male flowers. Fruits set on such plants are hybrids and give hybrid seed.

(d) **Use of gametocidal sprays:** There are certain chemicals known as gametocides which when sprayed before flowering induce complete male sterility. Thus without hand emasculation these plants can be pollinated to get hybrid seed. Cotton variety B-49 was made completely sterile by spraying 1.5 per cent solution of a chemical, FW 450.

(e) **Use of self incompatibility:** Self incompatibility has been reported in over 300 plant species including important plants like tomato, rape seed, alfalfa, kale, rye, etc. Thompson (1964) has suggested triple cross hybrid seed (A × B) × (D × E) × (E) in Kale (Figure 9.3).

6. ACHIEVEMENTS

Heterosis has been the most exploited research finding of the 20th century in plant sciences. Farmers have utilised the yield potential of hybrid varieties to a great extent (Table 9.4)

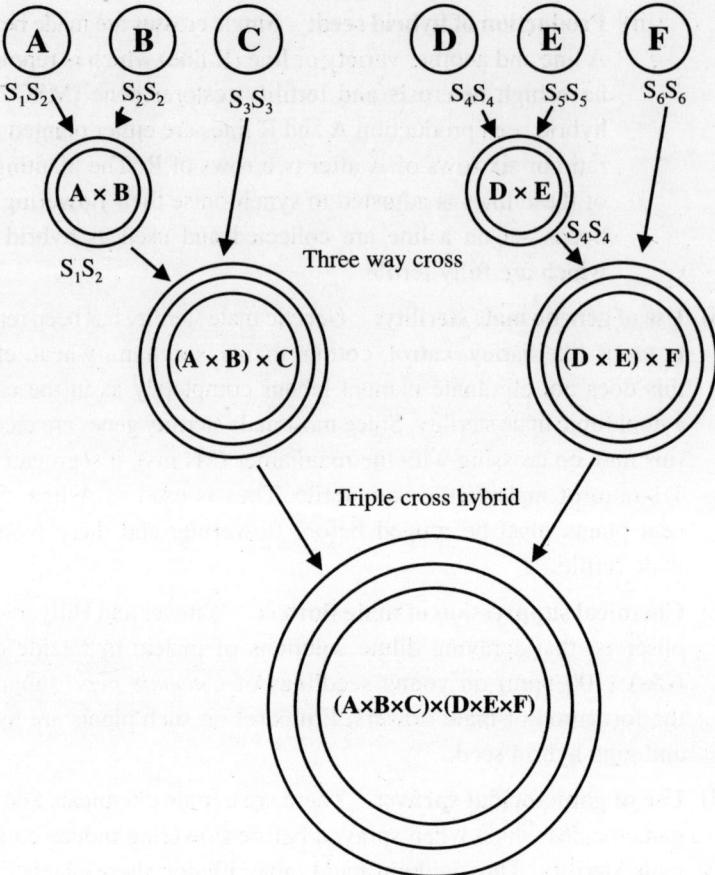

Figure 9.3   Triple cross Kale hybrid produced by using six inbreds each carrying a different incompatibility allele.

Many hybrid rice varieties (Wie yu 6, Sie yu 2, Shan yu 6) were first released in China, using cytoplasmic male sterility. More than 60 rice hybrids have been released until 201. Earlier rice breeders used WA source of cytoplasmic genetic sterility (cms lines) but now a number local cms (A) lines like Pusa CRMS 31A, CRMS 32A, TNAUCMS 2A, Patna CMS 1A, Pusa 6A, UPRI 95-17A etc have been developed and used for developing popular rice hybrids. Hybrid wheat is being attempted for production on a commercial scale using cms as well as chemical sterilants. Hybrid pigeon pea has been developed by ICRISAT Hyderabad. The future would open for the avenue of hybrid varieties in other crops also for breaking the yield barrier.

Table 9.4 Released varieties in India using heterosis (more examples in Chapter 14)

Crop	Hybrid variety
1. Bajra (Pearl millet)	HB-1 to HB-5, PHB-10, PHB-14, MBH-110 (for details please see Table 14.4)
2. Bottle gourd	Pusa Manjari (round), Pusa Meghdoot (long). Pusa Santushti
3. Brinjal	Vijay, Pusa Sankar 9, Pusa sankar
4. Castor	Gujarat Castor Hybrid-3 (TPS-10 R x Junagarh-1
5. Chillies	Chamatkar
6. Cotton	H-4, Varalaxmi (Lakshmi G. *hirsutam*) × SB. 2889 (*G. barbadense*), CBS 156, DH-7
7. Maize: Hybrids	Ganga 1, Himalayan 123,
Top cross	Ganga safed 2, Hi starch
Composite	Tarun, Vikas, Kisan, Jawahar (for details see Table 14.2)
8. Mango	Pusa Shreshta, Pusa Pratibha, and Pusa Arinina = (Amrapali x Sensation); Mallika and Amrapali = (Neelum x Dussehri)
9. Pigeonpea	ICPH-8
10. Potato	All varieties are heterozygous (see Table 14.6)
11. Rice	PRH 10, APRH-1, APRH-2, MGR-1, KRH-2, CNRH-3, Pant Sankar Dhan-1, PHB71, TNAU Rice Hybrid Co 4, CRHR-32, Sahyadri 4, JRH-5, Indira Sona, HKRH-1, NSD-2, Narendra Usar Sankar Dhan 3, PSD-3,
12. Safflower	DSH 129, MKH-11
13. Sorghum	CSH-1, CSH-2, CSH-3, CSH-4, CSH-5, CSH-6, CSH-7, CSH-8, CSH 26, CSH-27 (for details see Table 14.5).
14. Sunflower	BSH-1, DRSH-1,
15. Tea	TV 17, TV 24, Nandadevi, st. 397 (for details see Table 14.9)
16. Tobacco	GTH 1
17. Tomato	Pusa Sankar 8, Pusa Divya

7. USEFUL REFERENCES

Ashby, E. 1930. Studies in the inheritance of physiological characters I. A physiological investigation of the nature of hybrid vigour in maize. Ann Bot. 46: 1007-1032.

Davenport, C. B. 1908. Degeneration, albino and inbreeding. Science, 28: 454-455.

Dhawan, N.L. and Paliwal, R.L. 1964. The role of cytoplasm in the manifestation of heterosis. Maize Gent. Coop. Newsl.. 38: 70-71.

East, E. M. 1936. Heterosis. Genetics. 21: 375-397.

Gowen, J.W. (ed.) 1052. Heterosis. Ames. Iowa: Iowa State College, Press.

Jones, D.F. 1917. Dominance of linked factors as a means of accounting for heterosis. Genetics 2: 466-79.

Khanna, K. R. Chaudhary, R. C. and Chandra Prabha. 1971. A high yielding hybrid tomato. Plant Science 3: 36-38.

Rai, B. 1979. *Heterosis Breeding Delhi*, Agro. Bio. Publishers, New Delhi.

Srivastva, H.K. 1972. Mitochondrial complementation and hybrid vigour Ind. J Gent. Pl. Br. 32 : 215-228.

Thakur, B.J. and Seth, D.S. 1955. Commercial production of hybrid seed in Bombay state. 6th Conf. On Cotton Growing Problems, I.C.C.C., Bombay, pp. 9-12.

Mutation Breeding

1. MUTATION

"Mutations are sudden heritable changes in the genotype of an organism."

De Varies (1900) for the first time used the term mutation (*mutare* in Latin means "to change") for the appearance of new types in evening primrose (*Oenothera*) plant. In 1901 he gave the idea of producing mutations artificially for use in plant breeding. Alberto Pirovano (1922) was the first to use X-rays and ultra-violet rays for inducing mutations in plants but his work remained unknown. Therefore Muller (1927) became the first known to induce mutation in *Drosophila* fly using X-rays. For this work he received the Nobel Prize. Though Stadler (1928) had started working simultaneously on barley and maize and induced mutations yet he was late in reporting mutation due to longer life cycle of these plants compared to *Drosophila*. Mutation may be gene mutation; the result of change in the nuclear gene, plasma gene (cytoplasm e.g. chloroplast, mitochondria) or chromosomal mutation, change in chromosome number, structure or chromosomal aberrations of various types.

2. CLASSIFICATION OF MUTATION

The process through which genetic changes get induced is called mutation and the changed (mutated) individual is called mutant. Mutants have variously been classified as spontaneous and induced, natural and artificial based on their origin; germinal and somatic based on the tissue involved; chromosomal, genic and cytoplasmic, etc. Following classification put forth by Gaul (1961) is followed currently.

2.1 Macromutations

These are large mutations and can be recognised on a single plant basis. These are also of two types:

(i) **Trans-specific:**   Mutants, which resemble to a different species or even family, though these rarely occur, are called trans-specific. For example, the speltoid mutant of cultivated wheat (*Triticum aestivum*) resembles *T. spelta*, a wild wheat. The 'node-less' mutant of barley, where all nodes are crowded above root and only the last internode elongates to form the stem, resembles members of the family Cyperaceae.

(ii) **Intra-specific:**   These macro-mutants resemble the same species, for example, dwarf mutants of rice and wheat, etc.

2.2 Micromutations

These are mutations with small effects and can be recognised only when a group of 30 or more mutant plants are compared with a normal one. Micromutants differ with normals only quantitatively and supposed due to mutation governing by polygenes. For example, mutants with larger or smaller grains or higher yield etc. Thus micromutations are more important for direct use in plant breeding.

Point mutation is another term often used to designate gene mutation but it comprises of group of changes at individual locus (point) including micro-structural changes, micro-deficiencies and gene mutation.

Somatic mutation refers to mutants appearing in vegetative part in M_1 generation. It also refers to 'bud-Sport' in the case of vegetatively propagated plants. This may occur either due to dominant mutation (aa $\rightarrow$ Aa), recessive mutation in a heterozygote (Aa $\rightarrow$ aa), removal of epistatic factor or chromosomal aberrations.

2.3 Spontaneous Mutations

These are naturally occurring mutations, which arise automatically. They arise in nature continuously without any human control and create variability, which forms the basis of conventional crop breeding methods. Their frequency is extremely low (one in a million). The varieties of pea, which Mendel used in his crosses, arose by spontaneous mutation only. A number of factors cause spontaneous mutation. These are:

(i) **Genetic Constitution:** Spontaneous chromosome structural changes occur in influence of hybridity, polyploidy, or due to unstable chromosomes, etc. For example, plants with telocentric chromosomes in maize and wheat result into misdivision, non-disjunction and isochromosomes. Some genes like 'sticky (3 st)' in maize and wheat induce stickiness in chromosomes at mitosis and meiosis resulting in abnormal plants. The 'controlling elements' of maize increase the rate of mutation manifold when present in a particular individual.

(ii) **Physiologic conditions:** Many physiological factors, particularly those associated with aging, are responsible for spontaneous mutation. With the increase in the age of seed, germination decreases and chromosomal aberrations and mutation frequency increases.

(iii) **Nutrition:** In dog flower (*Antirrhinum majus*) deficiency of sulphur, phosphorous or Nitrogen increases spontaneous mutability. Environmentally induced heritable changes (transmutations) have been reported in linseed and tobacco also (Durrant, 1962).

(iv) **Temperature:** High temperatures induce mutations in nature. In Crepis, the mutation rate is abnormally high in places with 50°C temperature.

(v) **Naturally Occurring Mutagen:** Various agents known to induce mutations occur in nature in various forms. Certain place like the monazite belt of Kerala is known to have very high level of back ground radiation due to the presence of radioactive isotopes in soil. Besides, many chemical mutagens are present in the soil.

2.4 Induced Mutations

Contrary to spontaneous mutations, these are induced by using various agents called mutagen. Mutagens are the agents which induce mutation.

3. MUTAGEN AND ITS CLASSIFICATION

Physical or chemical agents who cause mutation are known as mutagens or mutagenic agents. There is a long list of mutagens now, which are classified as given in Figure 10.1.

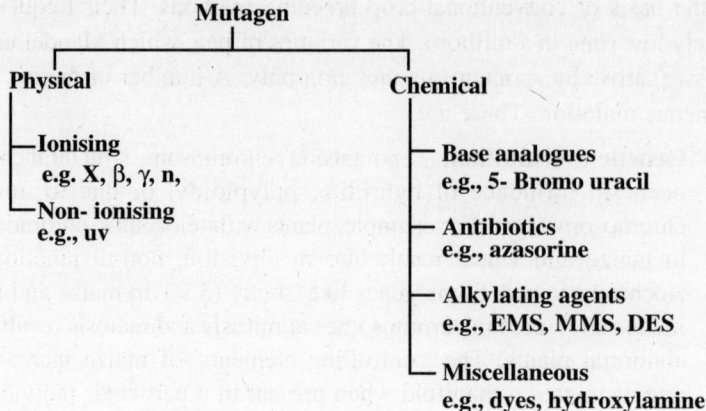

Figure 10.1 Classification of mutagen.

3.1 Physical Mutagen

The ultimate source of all physical mutagens is the atom. It is, therefore, important to have a clear idea of atomic structure. An atom consists of a nucleus and negatively charged electrons revolving in orbit. The atomic nucleus consists of positively charged protons and neutral neutrons (Figure 10.2). The entire weight of the atom lies in the nucleus, which in a stable atom consists of an equal number of protons and neutrons. The atom contains a high amount of bound energy. An unstable atom tends to give off energy or particles, thereby achieving stability. This is known as nuclear decay. Such unstable atoms are known as radioactive isotopes or radioisotopes and are defined as atoms of the same element having different weight. *The energetic atomic particles or electromagnetic waves accompanying nuclear decay are known as radiation.* The treatment of organism or plants with radiation is called irradiation.

Physical mutagens are further classified either as electromagnetic (e.g., X, UV, γ rays) and corpuscular (e.g., β, α, neutrons) or as non-ionising and ionising types. Radiations in the form of high energy short waves, which cause electric and magnetic disturbances in the matter through which they pass, are known as electromagnetic radiations. Radio waves, infrared, visible light, ultra-violet (UV), X and gamma (γ) rays (Figure 10.3) are electromagnetic but only the latter three are known to be mutagenic. Radiation in form of high-energy particles, which can transfer their kinetic energy to any matter through which they pass, is known as particulate or corpuscular radiation. For example all the particulate (α, β, neutrons) and non-particulate (X-rays, γ rays) radiations are ionising radiation. There is another kind of radiation called non-ionising radiation e.g. UV which does not ionise the substance through which it passes.

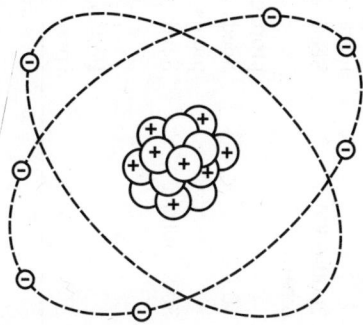

Figure 10.2 Structure of oxygen atom which consists of 8 protons and 8 neutrons in the nucleus and 8 revolving electrons.

3.1.1 *Non-ionising*

Ultra-violet (UV) rays are non-ionising radiations with wave length between 1000 to 4000 A, i.e. of very long wave lengths and thus have very little energy and penetrating power. They cause simply excitation through fluorescence, which results in increased chemical reactivity inside the tissue. UV is not efficient in producing mutants in organs other than pollen grains. The source of UV is the mercury vapour lamp and tube especially designed for UV rays. Chief characters of various physical mutagens are given in Table 10.1.

3.1.2 *Ionising*

(i) **Alpha (α) Particles:** Alpha particles are radioactive particles made up of two protons and two neutrons with positive charge. Thus in a sense these particles are naked helium nucleus. They are emitted chiefly by the isotopes of heavier elements and are strongly ionising. Being positively charged, they are liable to be stopped by negative charges to tissues and so have less penetrating power.

(ii) **Beta (β) Particles:** They have very less ionising power but more penetrating power than alpha particles. There is considerable variation in their energy and penetrating power, which depends on the number of the shell from where the electron is released. Beta particles are high-speed electrons with negative charge and are usually obtained from

Table 10.1 Chief characters of various physical mutagens

Mutagen	Source	Description	Penetration into tissue	Ionizing particle	No. of ion pairs/ micron of tissue	Material treated
X-rays	X-rays machine	Electromagnetic, ionizing	Few to many cm	Electron	100	Seed, seedlings, small propagule
Gamma rays	Radio isotopes (^{60}Co, ^{137}Cs)	Electromagnetic, ionizing	Many cm	Electron	8 any	Seed, seedlings propagule or plants in gamma garden
Neutrons: fast, slow, thermal	Nuclear reactor, accelerators	Neutral, ionising	Many cm	Proton	2000-6000 other	Seed, cuttings & propagules
Beta particles	Radio isotopes	An electron, densely ionising	Up to several mm	Electron	10	Seeds & plants
Alpha particles	Radio isotope	A Helium nucleus	Small fraction of a mm	He nucleus	3000-5000	Seed
Ultra violet	UV lamp	Electromagnetic, non-ionizing	Fraction of a mm	–	–	Pollen, spore

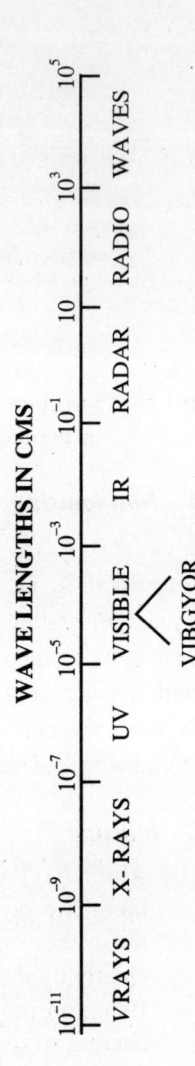

WAVE LENGTHS IN CMS

10^{-11}	10^{-9}	10^{-7}	10^{-5}	10^{-3}	10^{-1}	10	10^{3}	10^{5}
V RAYS	X- RAYS	UV	VISIBLE	IR	RADAR	RADIO	WAVES	

VIBGYOR

Figure 10.3 Wave lengths (in cm) of various radiations.

^{3}H, ^{32}P and ^{35}S radioisotopes. The material to be treated is soaked in the isotope solution. Injections in the tissue are also made. Seeds once soaked in isotope solution retain it and as a result radiation is received by somatic as well as reproductive cells of the developing plant.

(iii) **Neutrons (n):** Thermal (0.25 MeV) and fast (0.5 to 14.0 MeV) neutrons of wide energy range may be obtained from atomic reactors where ^{235}U is going under nuclear fission. Problems in dosimetry have created confusion in results otherwise neutrons are very effective, particularly in vegetatively propagated crops. They penetrate quite deep in the tissue as they are neutral and are not stopped by negative or positive charges of tissue.

(iv) **X-rays:** X-rays were first discovered by Roentgen in 1895. They have a wave length of 10-11 cm (hard X-rays) to 10-7 cm (soft) and originate from outside the atomic nucleus. X-ray machines, which have cathode tube, produce X-rays of varying wave lengths depending on the voltage, target, filter, thickness and composition of tube wall. The cathode tube consists of cathode, anode and target sealed in vacuum. The electrons are electrically accelerated and then stopped on a target made of tungsten or molybdenum. In accordance with electromagnetic theory, radiation is emitted, known as 'Bremsstrahlung'. But such radiations produce X-rays of variable wave lengths. X-rays with short wave length, known as hard X-rays, are more penetrating, compared to soft X-rays. By using appropriate filters X-rays are made more homogeneous.

(v) **Gamma (γ) rays:** Gamma rays have shorter waves than X-rays and therefore have more energy and are more penetrating. Gamma rays are emitted by a number of radioisotopes but in mutation breeding ^{60}C or ^{137}Cs are commonly used. Broadly the gamma emitting facilities may be classified into two groups. In the first group are the acute types where the source is strong and per minute high dosage are emitted out. Such machines may be of vertical or horizontal loading types depending on the method of putting the material for irradiation. Now such facilities are being manufactured at the Bhabha Atomic Research Centre, Trombay.

For chronic treatments, which require slow irradiation over long periods, facility is created over wide areas known as 'gamma garden'. In India, the first gamma garden was established at Bose Research Institute, Calcutta, in 1959. The second one was established at the Nuclear Research Laboratory,

Indian Agricultural Research Institute, New Delhi, in 1960. It has an area of three acres divided into various zones surrounded by 12 feet high and 3 feet thick and mud walls. The plants for treatment are kept in concentric rings. The centre one receive the maximum and the outer ones receive the minimum radiation from a ^{60}Co source kept in the centre.

3.2 Chemical Mutagen

Auerbach was first to initiate work on chemical mutagens during World War I (1914-18). She reported (1941) radiomimatic properties, that is, similar mutagenic property of some chemicals as of radiations. Chemical mutagens may be classed into the following four categories (Auerbach, 1976).

(i) **Base Analogues:** True to their name, the chemicals in this group are analogues of four bases of the DNA molecule, for example, 5-Bromouracil and 5-Bromo-deoxyuridine analogues of Thymine and pair with Adenine, and 2-Amino Purine (analogue of Adenine pairs with Thymine and Cytosine).

(ii) **Antibiotics:** A number of antibiotics like azasorine, mitomycine C and streptonigrin possess chromosome breaking properties. Their usefulness is limited, however, in practical breeding.

(iii) **Alkylating Agents:** This is the most important group of chemical mutagens, which react with DNA and act by alkylation of phosphate and purine and pyrimidine bases. Alkylation is a process by which an alkyl radical replaces hydrogen atoms in a molecule. For example, Methyl Methane Sulphonate (MMS), Ethyl Methane Sulphonate (EMS), Ethylene Imine (EI), sulphur, mustard gas, epoxides, etc, produce alkylation.

(iv) **Miscellaneous:** Hydroxylamine, nitrous acid, dyes like acridines, alkaloids, etc, have mutagenic properties. Difference between chemical and physical mutagens is summarised in Table 10.2.

4. MUTAGENIC TREATMENT

When the entire dose is applied in the shortest time and once (usually in seconds to minutes), it is called Acute. For example, the acute treatments of gamma rays is done in gamma trans or gamma cells.

Table 10.2 Difference between physical and chemical mutagens in their use and effectiveness.

	Physical mutagen	Chemical mutagen
1.	More efficient in vegetatively propagated crops	More in seed propagated crops
2.	Costly in terms of equipment and laboratories	Very economic
3.	Safe for the user excepting isotopes	Unsafe as most chemicals are carcinogenic
4.	Chronic treatments easy	Chronic treatments difficult or even not possible in most cases
5.	Random distribution of chromosomal breaks	Localised
6.	Less chromosomal breaks and more recombinants.	More chromosomal fragments and less recombinants.
7.	More of inter chromosomal changes and large chromosomal mutation	More if intra chromosomal changes like point mutation, duplications, deficiencies and minute alterations
8.	Penetration in tissues is very deep barring UV	Penetration in tissue very limited.
9.	No delayed effect or later appearance of mutation except with isotopes	Delayed effect and thus late appearance of mutants as chemical may remain with tissue

4.1 Acute

When the entire dose, usually high, is applied in the shortest time and once (usually in seconds or minutes). For example, the acute treatments of gamma rays is done in a gamma trans or gamma cells in few minutes.

4.2 Chronic

When the entire dose is applied over long periods (weeks or months), usually throughout the life cycle of a plant. For example, chronic treatment with gamma rays is done in "Gamma Garden" of IARI New Delhi. This uses a Cobalt (Co^{60}) source contained in a lead casing and surrounded by high wall in 3 acres. Not much in use currently but this facility has produced many mutants.

4.3 Recurrent

When the mutagenic treatments are repeated in two or more subsequent generations, for example, in M_2, M_3, or M_4 after M_1, it is known as recurrent.

5. DOSIMETRY

The dose of X and gamma rays is measured in Roentgen (r) units which is defined as the quantity of radiation whose associated corpuscular emission per 0.001293 gm of air produces in air, ions carrying one *esu* of electricity per cc of air at NTP. The r is expressed in *mr* ($= 10 - {}^3r = 0.001$ r) and Kr (1000r). Other measurements are:

5.1 Radiation Absorbed Dose (RAD)

One *rad* corresponds to the absorption of energy of 100 ergs/gm of tissue.

5.2 Radiation Equivalent Physical (REP)

It has been introduced to measure the ionisation by X, β and *n* particles. It corresponds to the amount of any kind of radiation producing the same number of ion pairs or energy in tissue or water which are produced by one roentgen (r) of X or gamma radiation.

5.3 Curie (C)

Curie is defined as the activity of a radioactive isotope in which 3.7×10^{10} disintegration take place per second.

5.4 Lethal Dose 50 (LD$_{50}$)

This is the dose of a mutagen on which 50 per cent of the treated individuals get killed. This is an important parameter as doses close to LD$_{50}$ are reported more effective in inducing maximum mutation. The LD50 of some plants is given in Table 10.3.

Table 10.3 LD$_{50}$ range of some plants compared to 500 r of man

Dose in r	Crops
1000–2000 r	Canna rhizomes, grape cutting
2000–5000 r	Buds, apple scion, cassava, sugarcane, tea cutting
5000 r–10,000 r	Safflower, soybean, pea, urd bean, mung bean
10,000–15000 r	Poa, rye, *Lupinus*, maize, Petunia
15,000–20,000 r	Groundnut, oat, sweet potato
20,000–30,000 r	*Trifolium*, tobacco, wheat
30,000–40,000 r	Sorghum, cotton, rice
50,000–80,000 r	Linseed, radish, *Vigna*, barley
1,00,000 r	Mustards, *Sinapis*, *Populus*

6. MODE OF ACTION OF MUTAGENS

A. Physical Mutagen

When radiation passes through seed or any other plant propagule, it creates ionisation, excitation and fluorescence. Ultimately all the radiation energy in the tissue becomes degraded to heat energy. Barring UV which has only fluorescence effect, other radiations have fluorescence, excitation and ionisation effects. The total effect results into mutation or death of the organism. There are many hypotheses related to death of the organism and tissue but the following two explain the mutagenic mechanism of irradiation.

6.1 Activated Water Hypothesis

Dale (1940) developed this hypothesis on the basis that irradiated tissue has lots of water and it produces radicals. Thus OH, OH$_2$ and H radicals and H$_2$O$_2$ are produced, which cause irreversible oxidation of chromosome parts and, on collision with chromosome molecule, transfer energy to them which causes point mutations (Figure 10.4).

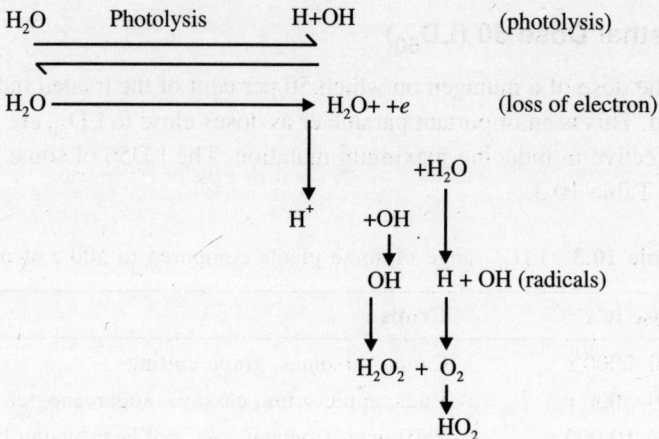

Figure 10.4   Pathway of activated water hypothesis.

6.2 Target Theory

Given by Crowther (1924) and developed by Delbruck, Timofeev-Ressovasky, Zimmer and Lea (1930-1940), this theory supposes the existence of target or sensitive area within cell, damage to which results in biological changes. Only those ionising particles, which hit on the target are effective; thus frequency of point mutation and chromosome break is directly proportional to dose.

B. Chemical Mutagens

6.3 Transition

Transition is the transfer of one purine (adenine, guanine) by another purine or one pyramidine (thymine, cytosine) by another in a DNA molecule.

   Out of four bases of DNA, adenine pairs with thymine and guanine with cytosine but in a mutated state a rare adenine can pair with cytosine creating an abnormal DNA molecule, which will mutate.

6.4 Transversion

Transversion is substitution of purine bases by pyramidine and vice versa and leads to mutation.

6.5 Deamination

Replacement of amino group by hydroxil group is called deamination. Deamination of cytosine, for example, produces uracil, which pairs as thymine. This type of defective pairing leads to mutation. Many of the mutagenic damages in cell can be repaired back but most mutagens inactivate the repair enzymes also.

7. PROCEDURE OF MUTATION BREEDING

Procedure of mutation breeding is depicted in Figure 10.5.

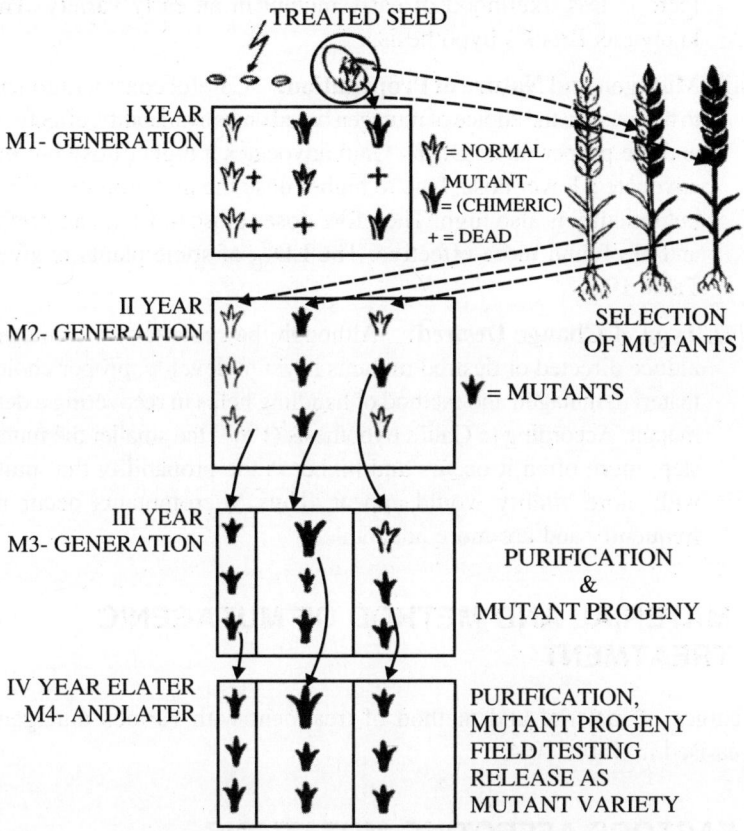

Figure 10.5 Procedure of mutation breeding.

7.1 Basic Considerations

(i) **Available Variability:** If the desired type of variability exists in nature it is better not to go for mutation breeding, but rather to use the available variability. Otherwise mutation breeding is a must.

(ii) **Material and Nature of Propagation:** The choice of mutagen, procedure of treatment and procedure of handling M_1, M_2 and later generations depends entirely on the type of material to be treated and the nature of its propagation such as self or cross pollinated, asexually propagated, etc care has to be taken while choosing material for the induction of micromutants. Dr. R. D. Brock (1964) has hypothesised that the mean of mutant population deviates in a direction, which is opposite to the previous selection history of parent material. Thus there is less likelihood of early mutant in an early variety. This is known as Brock's hypothesis.

(iii) **Mutagen and Nature of Propagation:** Careful consideration needs to be given on the choice of mutagen based on its availability, effectiveness and the proper dose. Dr. H. Gaul advocates a higher dose but others advocate a lower dose. Due to higher dose the mutation rate is higher but lethality is also high. Therefore doses close to LD_{50} are preferred and are found more effective. The LD_{50} of some plants is given in Table 10.3.

(iv) **Type of Change Desired:** Although the specificity of mutagen to induce directed or desired mutants is yet to develop, proper choice of material, mutagen and method of handling helps in recovering a desired mutant. According to Gaul's hypothesis (1967) the smaller the mutation step, more often it occurs and higher is the probability that mutants with more vitality would appear. Thus Micromutants occur more frequently and are more adaptable.

8. MATERIAL AND METHOD OF MUTAGENIC TREATMENT

The choice of material and method of treatment with various mutagen are summarised in Table 10.4.

9. FACTORS AFFECTING MUTATIONS

A number of factors like radiation parameters, biological factors, environmental factors, pre-and post-treatment factors influence mutagenic sensitivity and thereby affect the frequency of mutation. Genetic compositions of a plant play important role, like some varieties or species are more sensitive and produce many mutations. Diploids are more sensitive than polyploids. Low moisture content makes seeds more sensitive to radiation. Then there are various radio-protective chemicals which reduce the effect of radiation. Chemical scavengers like Sodium Thiosulphate increase mutagenic efficiency of chemical

Table 10.4 Method of treatment of plant propagule with mutagens

Propagule \ Mutagen	Ultra Violet (UV)	Alpha rays (α)	Beta (β) particles	X-ray rays (γ)	Gamma (n)	Neutron mutagen	Chemical
Seed	No	Soaked in isotope solution	Soaked in isotope solution	Exposed in X-ray machine	Exposed in gammatron	Exposed in generator	Soaked in solution
Tubers, bulbs, cuttings etc.	No	No	- do -	- do -	- do -	- do -	- do -
Seedling, specific stage of plant	No	No	- do -	- do -	- do -	- do -	- do -
Plant throughout its life cycle	No	No	- do - or through soil	No	Exposed in gamma garden	No	No
Pollen	Thin layer exposed	No	No	- do -	No	No	No

Figure 10.6 Growth inhibition in tomato seedlings from seeds irradiated with gamma rays; SC = control, S_1 = 5,000 r, S_2 = 10,000 r, S_3 = 20,000 r (Courtesy: Ph. D. Thesis 1969, R C Chaudhary).

mutagen. Presence or absence of oxygen also plays important role. For a detailed description of these factors see Konzak et al. (1965).

10. HANDLING OF TREATED MATERIAL

10.1 Self-pollinated Crops

A general scheme is given in Figure 10.5

(i) **First Treated Generation (M_1):**   Terms like R_1, X_1, etc, are also used but the use of M_1 is proper, more common and advocated to designate the first treated generation. After mutagenic treatment the plant propagule is planted in the field along with untreated control.

There are many common effects observable in M_1 itself like death of seedling and young plants, growth inhibition (Figure 10.6) morphological and developmental abnormalities (Figure 10.7) and heritable changes in qualitative and quantitative characters. The first three categories of effects normally disappear in the M_1 generation itself. The heritable changes, that is, mutations, only are of importance to breeders which may involve changes sin chromosomes, cytoplasm and in genes itself. Only few mutants are observed in the M_1 generation which are either as dominant, double recessive or due to removal or creation of epistatic or inhibitory effects. Therefore, only few M_1 plants are selected as mutants and the majority of them are promoted to M_2 as such.

Figure 10.7 Developmental abnormality in tomato seedlings of seed treated with 20,000r of gamma rays (Courtesy: Ph.D. Thesis 1969, R.C. Chaudhary)

(ii) **M_2 Generation:**   It is the most important generation as maximum number of mutants are recovered due to recombination and segregation in diploid species. Usually the ear-to-row method is used to grow M_2

but sometimes one to three grains are picked from each M_1 plant. A large population is essential for efficient recovery of mutants. Since mutations may get induced affecting morphologic character, yield and yield components, biochemical and physiological mutants, their recovery would depend on the efficiency on screening procedure in M_2 and later generations. Therefore M_2 and later generations should be thoroughly screened for type of mutants desired.

Recovery of mutant in M_2 is the most important issue. The embryo is a miniature plant where many leaves, etc, already remain differentiated. Mutation takes place usually in a few meristamatic cells of the seed; thus the seedling developing out of it becomes chimeric. Mutant tissue ending up in the formation of vegetative parts like leaves, etc, would be lost (Figure 10.8). For a mutant tissue of a treated seed to be included in tissues forming seed on M_1 plant, it has to pass through two sieves of selection pressures.

1. **Diplontic selection:**   The mutated tissue of the treated seed at the time of germination faces diplontic selection, that is the diploid normal tissue and diploid mutated tissue compete with each other during the vegetative stage and during ontogenetic differentiation of reproductive organs. If the mutant sector is large enough or is able to compete even otherwise, only then it can enter into the formation of reproductive parts.

2. **Haplontic selection:**   If the M_1 flower or inflorescence is still chimeric, mutant and normal type of pollen would be formed. Again on the stigma and in style both types of pollens would compete for fertilising the ovule, and thus haplontic selection would operate. If this time also the mutant pollen is able to compete in haplontic selection only, then the mutant would be recovered in M_2 or later generations.

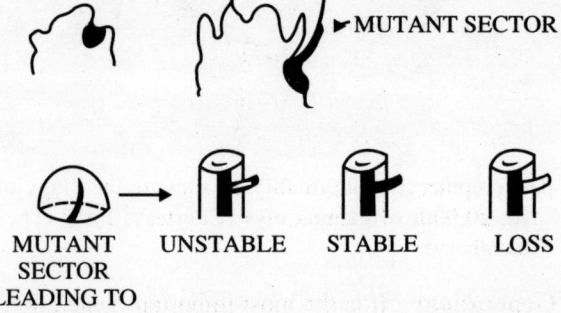

Figure 10.8 Fate of chimera: becomes stable if involves a bud or gets lost if dose not

(iii) **M₃ and Later Generations.**

In M_3, plant progenies of selected M_2 plants are grown and evaluated critically. Mostly eye-ball selection, that is, selection based on appeal to the eyes, is done in M_3. Progenies or single plants thus selected are grown in M_4 and later generations for further purification, evaluation and multiplication.

Usually by M_5 generation most mutants get purified and thus are evaluated in field trials along with parental variety and the standard check variety. If in repeated tests the mutant proves its superiority over standard check, it is released as a variety. Both macromutants and micromutants are evaluated for their potential.

10.2 Cross-pollinated Crops

Cross-pollinated crops have enough genetic variability already in nature hence have not been subjected to mutation breeding so much as self pollinated ones. Moreover these are heterozygous and open pollinated and thus detection of mutants is difficult. Of late, some attention is being given to these (Anon. 1976). The procedure of handling the M_2, M_3 generations is like self pollinated crops except that selfing is done to check out crossing.

10.3 Asexually Propagated Crop

It would be apparent from Chapter 3 that this group of plants either lack sexual reproduction or their sexual cycle is long. Therefore conventional breeding methods are difficult. Contrary to this, any mutant including somatic ones can be propagated and directly used. For detailed information see Anon. (1973).

(i) **Genetic and anatomical considerations:** Since such plants are heterozygous, any mutation from dominant to recessive may be detected and used. Even any phenotypic effect of chromosomal arrangements may also be utilised owing to the vegetative mode of propagation. The outer cell layers (three in most plants) possess certain degree of individuality during vegetative growth and differentiation. As mutations are single cell, even they will be limited to one particular cell layer. The mutated plant will become a chimera-ecto, meso or endo chimera depending on whether the epidermis or some lower tissue is changed.

The chimeric structure is usually quite constant on vegetative propagation but easily changed in several ways and hence is the most important factor in mutation breeding, before the mutated tissue forms a sectorial (mericlinal) distribution in the shoot developing from the treated bud. If the sector is narrow, for example, after irradiating the highly differentiated bud, it may be lost (see Figure 10.7) during growth and propagation. This is why less differentiated primordia are treated. Diplontic selection, that is, competition between mutated and normal cell, which is normally in favour of the latter, reduces the recovery of mutants. Formation of chimera is avoided by irradiating the youngest possible stage of bud. Selection of suitable propagation methods like budding, stooling, etc, and pruning the shoots of mericlinal chimera are also followed. Another consideration is that the material to be treated should be free from viruses.

(ii) **Mutagenic treatment:**   Physical mutagens like X-ray, gamma rays and neutrons have been more effective than chemical mutagens. This is because physical mutagens penetrate deep and produce more chromosomal re-arrangements. Therefore X and γ rays and neutrons are used more. Acute exposures are generally more effective than chronic ones.

(iii) **Handling the treated material:**   Chimerism is associated with mutated sector to grow by pruning the normal or other vegetative propagate the basal buds of primary shoot emerging from the treated bud. The other method could be to purify the chimera through various propagational means like grafting, budding, etc. The main consideration is avoiding diplontic selection and loss of mutant sector in the chimera.

(iv) **Time taken:**   The time taken to develop a superior variety depends on many factors like type of material, mutagen, procedure, etc. If a suitable mutant is obtained in M_2 itself, the time taken is very less, as compared to conventional methods. In vegetatively propagated material, the time taken is definitely much lesser as compared to conventional breeding methods.

11. APPLICATIONS AND ACHIEVEMENTS OF MUTATION BREEDING

11.1   Use of Point Mutation

A.   Self-pollinated Crops

(i) **Defect elimination:**   Sonora 64 variety of wheat had red colour grain. This was irradiated with gamma ray and Sharbati Sonora was produced with amber colour and acceptable grain colour (see Table

14.3). Semi-dwarf rice variety Jagannath was produced from tall T141 by gamma rays (Table 14.1).

(ii) **Improvement through micromutation:**   A higher yielding mutant NC4 of groundnut variety NC4 was released though it looks closer to its parents.

(iii) **Use of mutants in hybridisation:**   Mutants, if not agronomically superior, are intercrossed or crossed with another variety to produce superior varieties. Several non-productive mutants of groundnut were intercrossed to produce high yielding varieties by Patil (1979).

B.   Cross-pollinated Crops

(i)   To increase variability

(ii) To exploit mutated genes in mutational heterosis

(iii) To induce male sterility for use in hybrid seed production.

C.   Asexual Plants

Point mutations usually result in sports or chimeras, which may be purified and multiplied and, if superior, can make a variety. Sugar cane varieties TS-1 and TS-2 are mutants of Co 419 (see Figure 8.3). Sugarcane varieties Co8152 and Co8153 are mutants of Co 527 and Co775 respectively.

11.2   Use of Chromosome Mutations:

(i)   Use of polyploids has been done in crop improvement. Polyploids have been induced in many crop plants and used (see Chapter 11).

(ii) Use of translocations for transferring characters from related species and genera chromosome segments of *Aegilops, Secale, Agropyron* carrying rust resistance genes have been translocated to wheat by using X-rays (see Chapter 11).

(iii) Use of translocations with known breakage points for the production of directed-duplication of a segment of chromosome 6 of barley carrying gene 'Orange lemma' which is associated with amylase activity, improves the malting quality of barely.

(iv) Diploidisation of polyploids is another way out in plant breeding. Wheat is hexaploid but behaves as diploid due to gene Ph located on chromosome-5 B. This information is used to make other polyploids behave as diploids.

11.3   Use in Special Breeding Purpose

(i) **Breaking Undesirable Linkage:**   Resistance to victoria blight in oats was linked with susceptibility for crown rust, which was broken using mutagen (See Chapter 12).

(ii) **Production of Haploids:**   Haploids may be produced using X-rayed pollen for pollination. Unfertilised eggs develop into haploid which are variously used (see Chapter 11).

(iii) **Increase or Decrease of Chiasma Frequency:**   Chiasma frequency is an inherited character and has an important bearing on recombination in segregating generations. Chiasma frequency can be increased or decreased with mutations.

(iv) **Production of Transitory Sexuality in Apomicts:**   Apomictic plants breed like asexually propagated ones and hence are difficult to breed. But transitory sexuality can be induced and better apomictic types can be selected as in Poa (see Fig. 8.2) after crossing the various sexual types.

(v) **Reduction of Incompatibility in Wide Crosses:**   Through use of irradiated pollen, interspecific crosses in Avena and Nicotiana genera have been made. The actual mechanism is not known but irradiation induces better tube growth and produces more auxin in style.

(vi) **Studies of genetic, morphologic and biochemical processes in plants:**   Mutants in the same genetic background or at same complex locus are used for such studies.

(vii) **De *novo* (New) Creation of Genetic Variability:**   When the desired character does not exist, mutations are used. For example, ergot resistance in bajra does not exist and hence mutation breeding is being looked for.

11.4   Varieties Developed by Mutation Breeding

The detail list is available on the websites: **www.mvd.iaea.org/MVD/default.htm; www.mvgs.iaea.org./AboutMutantVarieties.aspx)**

Many varieties have been released through mutation breeding (Table 10.5, Figure 10.9) globally. Most mutant varieties have been developed in cereals followed by ornamental plants (flowers).

Table 10.5. Mutant varieties released up to 2013 in all countries of the world. (Source: Dr. Stephen Nielsen, 2013: www.mvgs.iaea.org./ AboutMutantVarieties.aspx)

Group of crops	No. of released varieties
1. Cereals (wheat, rice, barley etc)	1,589
2. Flowers	642
3. Legumes	492
4. Oilseeds	110
5. Others	385
Total	**3,218**

Figure 10.9 Mutant variety Aruna of castor (Courtesy: Dr. N.G.P. Rao)

More than 332 varieties (Table 10.6) of food crops, medicinal plants, flowers etc have been released in India by the year 2013. Some prominent and recent ones are **barley** (K-257, Karan 265, Rajkiran), **Urd bean-black gram** (DU-1, TU 94-2, Vamban-2), **bougainvillea** (Arjun, Los Banos Variegata), **castor** (Aruna), **brinjal** (PKM-1), **chickpea** (Ajay, Atul, Girnar, BGM 547, **chrysanthemum** (Alankar, Anamika, Basant, Kanak, Anamika, Jugnu, Shabnam, Sonali, Subarna, Surekha Yellow), **cotton** (Indore-2, Khandwa-2, MCU-10, MCU-7, Pusa Ageti, Rashmi), **cowpea** (COCP702, Amba, Shreshta), **French bean** (Pusa Parvati), **groundnut** (GPBD 5, Mutant 28-2, TG1, TAG-24, TDG-39, TG 39), **jute** (JRO 514, JRO 412, Savitri), **lady's finger** (Anjitha), **lemon**

Table 10.6   Mutant varieties of field crops, fruits, vegetable and ornamental crops released in India till 2013 (Compiled from: www.mvgs.iaea.org./ AboutMutantVarieties.aspx)

Name of crop	Mutant varieties	Name of crop	Mutant varieties
Barley	13	Mung bean (green gram)	17
Bean	1	Papaya	1
Bitter gourd	1	Pea	1
Bottle gourd	1	Pearl millet	5
Bougainvillea	10	Pigeon pea	7
Brassica (mustards)	5	Poppy	1
Brinjal (egg plant)	1	Portulaca (Moss rose)	14
Castor	2	Rapeseed	1
Chickpea	6	Rice	60
Chillies	1	Ridge gourd	1
Chrysanthemum	49	Rose	14
Clover	1	Sesame	3
Coleus	1	Sorghum	1
Cotton	10	Soybean	7
Cow pea	9	Sugarcane	5
Dahlia	10	Sunflower	1
Gladiolus (Gladioli)	2	Tobacco	1
Groundnut	26	Tomato	4
Hibiscus (Shoe flower)	2	Tuberose (Polianthes)	2
Jute	6	Turmeric	2
Lady's finger	2	Urd bean (black gram)	8
Lemon grass (Citronella)	6	Wheat	2
Lentil	3	Ziziphus (Ber)	1
Mulberry	1	Others	3
Total			**332**

grass (Niranjan), **lentil** (Rajendra Masoor 1), **green gram - mungbean** (BM 4, LGG 450, Pant Moong 2, TJM 3, TJM 2000-2, TGM 96-2), mustard (Kranti, TPM-1), p**apaya** (Pusa Nanha), **pea** (Hans), **pearl millet** (NHB4), **pigeon**

pea (TJT 501), Trombay Visakha 1), **rapeseed** (NUDB-38), **rice** (Anaswara, CRM207-1, Early Samba, Gautam, Jagannath, Lunisree, Malviya Dhan 36, Pusa -NR-519, Pusa Jaldi Dhan13, IIT 60, K84, HM 95, Ramchandi, Sattari, Tapswani), **rose** (Light Pink, Saroda, Sukumari, Twinkle), **sugarcane** (Co8152, Co8153), **sesame** (Uma, Usha), **soybean** (Birsa Soybean 1, MACS-450, TAMS 38), **tobacco** (Jaysri), **tomato** (Pusa Lal Meeruti, S 12), **turmeric** (BSR-1), **wheat** (NP 836, Pusa Lerma, Sharbati Sonora) etc.

In the varieties released through mutation breeding almost every character has been improved as depicted in Table 10.7.

Table 10.7   Characters improved through induced mutation in crops. (Sigurbjorasson and Micke, 1974)

Improved characters*	Cereals	Legumes	Others	Total
Higher yield	27	10	10	47
Lodging resistance	23	3		26
Disease resistance	13	9	2	24
Early maturity	19	9	8	36
Short stem	14	2		16
Quality	13	3	11	27
High protein	2	2		4
Shattering resistance		2		2
Easier harvesting	1	2		3

* Other improved characters include cold tolerance, improved plant type, grain colour and weight, dormancy, drought tolerance, high lysine.

12. LIMITATIONS

12.1 Unpredictability

There is no control over the mutation process or rate. Hence most of the mutants are useless and only a few are of economic importance. Mutation is still a hit and miss process.

12.2 Low Frequency

Mutation rate is very low and to recover any useful mutant a large population in M2 and m3 generations need to be handled. Even at LD_{50}, mutation rate is very low.

12.3 Special Facility

Special facilities and equipment are needed which are not normally available at every breeding station.

12.4 Unrepeatability

Since the mutation process is not every well understood, no control can be exercised on the outcome. Unless directed mutagenesis is well understood, it is difficult to obtain the same or desired mutants every time.

12.5 Lethality

Lethality of even otherwise beneficial mutants poses problems in further multiplication of the material, though this is only occasional.

13. USEFUL REFERENCES

Anon. 1961. Mutation and Plant Breeding. NAS, NRC.

---1965. The use of Induce Mutations in plant Breeding. Oxford; Pergamon Press.

---1966. Mutations in Plant Breeding. I. Vienna: IAEA.

---1968. Mutations in plant Breeding. II. Vienna: IAEA.

---1973. Induced Mutations in Vegetatively Propagated Plants. Vienna: IAEA.

---1976. Induced Mutations in Cross Breeding. Vienna: IAEA.

---1976 to 1980. Mutations Breeding Newsletters. Vienna: IAEA.

Anon. 1997. Manual on Mutation Breeding. Technical Report Series No. 119, IAEA, Vienna, Austria.

Auerbach, C. 1976. *Mutation Research*. Chapman and Hall, London.

Bhatia, C. R. 1996. Economic impact of mutant varieties.. In: Plant Mutation Breeding for Crop Improvement. IAEA, Vienna.

Brock R. D. 1957. Mutation Plant breeding, Jour. Aust. Inst. Agrl. Sci. 23:39-50.

Brunner, H. 1991. Methods of Induction of Mutations. In: A. K. Mandal, P. K. Ganguli and S. P. Banerjee (Eds.) Advances in Plant Breeding. SBS Publishers, Delhi.

Chaudhary, R. C. 1970. Cytogenetics and breeding of tomato. Ph.D. thesis, Agra University, Agra.

Chopra, V. L. (Ed.) 2008. *Plant Breeding: Theory and Practice*, Oxford & IBH Publishing Co. Pvt. Ltd., New Delhi.

Durrant, R. 1962. Environmental induction of heritable change in *Linum*. Heredity 17:26-61.

Gaul, H. 1964. Mutations in plant breeding. Bot. Rev. 4: 155-232.

Hollaender, A. (Ed.) 1954. *Radiation Biology*. McGraw-Hill, New York.

Konzak, C.F., Nilan, R.A., Foese-Gertzen, E.E. and Foster, R.J. 1965. Factors affecting the biological action of mutagens. Proc. Symp. Induction of Mutations and Mutation Process, Prague.

Prasad, A. B. (Ed), 1986. *Mutagenesis Basic and Applied*. Print House, India, Lucknow.

Rai, M. and Prasanna, B. M. 200. Transgenics in Agriculture, ICAR, New Delhi.

Rakshit, R. C. (Ed.) 1989. *Heterosis Breeding*. Bidhan Chandra Agriculture University, Kalyani, West Bengal, India.

Ram, H. H. and Singh H. G. 1994. *Crop Breeding and Genetics*. Kalyani Publishers , Ludhiana, Punjab.

Singh, B. D. 1998. *Biotechnology*. Kalynai Publishers, Ludhiana.

Verma, M. M., Virk, D. S., Chahal, G. S. and Dhillon, B. S. (eds.) *Heterosis Breeding in Crop Plants*: Theory and Practice. Crop Improvement Society of India, PAU Ludhiana, India.

Sigurbjorasson, B. and Micke, A. 1974. Philosophy and accomplishments of mutation breeding. in: Polyploidy and Induced Mutations in Plant Breeding. Vienna: IAEA.

Ploidy Breeding

1. DEFINITIONS

The chromosome number of any species, whether diploid or polyploid is designated as $2n$. In a somatic cell $2n$ is the number of chromosome that exists whereas in normal gamete (pollen or ovule) the number is half of this, that is, n and thus called as *gametic number* (n). The basic chromosome number (x) of a species which often corresponds to the *gametic number* (n) in a true diploid is known as *genome* or full chromosome complement. Each chromosome of a genome exists in a pair form or homologues in the somatic cells. Different chromosomes of a genome are distinct from each other in morphology and gene content. Chromosomes of a single genome do not show any tendency of pairing with another member of that genome. The number of chromosomes in a somatic cell constitutes the somatic number ($2n$) in a diploid plant. For example, barley has 7 chromosomes in its pollen, the basic or gametic number being 7, and 7 pairs or 14 chromosomes ($2n$) in the somatic cell. The number of times the basic number is repeated in the somatic number is called ploidy. In barley the basic number or gametic number ($x = n$) is repeated 2 times in its somatic number and thus barley is a diploid (2n). In a polyploid plant somatic number is more than $2n$.

2. CLASSIFICATION OF PLOIDY

Based on levels of ploidy two groups of plants exist: (a) whose somatic chromosome number is an exact multiple of the basic number, and (b) whose somatic number is not an exact multiple. The plants of the first group are called euploids and those of the second group called aneuploids. The euploids and aneuploids are classified further (Table 11.1) taking a diploid with 6 somatic chromosomes A_1A_1 A_2A_2 A_3A_3. The level of ploidy and its type in a crop plant has a very important bearing on the choice of plant breeding methods and procedures.

2.1 Aneuploid

(Aneu = not, ploid = multiple). If the somatic chromosome number is not exact multiple of the basic number, it is known as an aneuploid. If both the members of a chromosome pair miss, it is known as nullisomic. Similarly, if one or both members of a chromosome are present in addition, it is known as trisomic and tetrasomic respectively. Many other types of combinations are possible but they do not have much importance in breeding.

2.2 Effect of Aneuploidy

Usually the aneuploids have abnormal or subnormal morphology and are less vigorous due to an imbalance in the number of chromosome and associated physiological disturbance. They do not have any agricultural importance as varieties. Monosomics do not survive in diploids like barley, but in polyploid species like wheat (hexaploid but behaving as diploid) and tobacco (allotetraploid) they do survive. Trisomics, however, survive in both diploids as well as polyploids. Datura, maize and tomato are classical examples of trisomics.

2.3 Application of Aneuploids in Crop Breeding

1. **Effect of individual chromosome:** By studying series of aneuploids for all the chromosome one by one, effect of individual chromosome on phenotype can be seen.

2. **Locating linkage group:** By studying secondary or tertiary trisomics, the gene may be located on a particular arm of the chromosome.

3. **Genome relationship:** Aneuploids provide opportunity for chromosome of one genome to other if they are homoeologous. For example a wheat plant with 2A tetrasomic & 2B nullisomic would appear phenotypically normal as A genome would compensate for the loss of B genome.

Table 11.1. Classification of plants based on ploidy level, where gametic number = basic number = 3(A1A2A3); diploid somatic chromosome = A1A1 A2A2 A3A3; and A,B,D, are three different genomes

Type of ploidy	Somatic formula	Somatic chromosome
1. Aneuploids:		
Nullisomic	$2n - 2$	$A_1A_1\ A_2A_2$
Monosomic	$2n - 1$	$A_1A_1\ A_2A_2\ A_3$
Double monosomic	$2n - 1 - 1$	$A_1A_1\ A_2A_3$
Trisomic	$2n + 1$	$A_1A_1\ A_2A_2\ A_3A_3A_3$
Double trisomic	$2n + 1 + 1$	$A_1A_1\ A_2A_2\ A_2\ A_3A_3A_3$
Monosomic-trisomic	$2n - 1 + 1$	$A_1A_1\ A_2\ A_3A_3A_3$
Tetrasomic	$2n + 2$	$A_1A_1\ A_2\ A_2A_3A_3$
2. Euploids:		
(a) Autoploids:		
Monoploid / Haploid	x	$A_1A_2A_3$
Diploid	$2x$	$A_1A_1\ A_2A_2\ A_3A_3$
Triploid	$3x$	$A_1A_1A_1\ A_2A_2A_2\ A_3A_3A_3$
Tetraploid	$4x$	$A_1A_1A_1A_1\ A_2A_2A_2A_2$ $A_3A_3A_3A_3$
(b) Alloploids:		
Triploid	$x + x' + x''$	ABD
Tetraploid (Amphidiploid)	$2x + 2x'$	AADD
Hexaploid	$2x + 2x' + 2x''$	AABBDD

3. EUPLOID

(*Eu* = good, *ploid* = multiple). Those species with the exact multiple of the basic chromosome number come in this category. Depending on the multiples of the basic number present, euploids may be monoploid (x), diploid ($2x$), triploid ($3x$) tetraploid ($4x$), hexaploid ($6x$) and so on. The term haploid refers to individuals having gametic number may be from a diploid or polyploid. The haploid from a true diploid would be a monoploid. For example, barley haploid would be a monoploid but a potato haploid would be a diploid since potato is tetraploid. Hybridisation among these types leads to intermediate euploids of 3n, 5n and so on. Species with chromosome numbers above the diploid level are collectively called polyploid. Polyploids having the same genome are termed autopolyploids, and those with different genomes are termed allopolyploids. The allopolyploids arise by the hybridisation of two species having different genomes, followed by doubling of chromosomes. For example, bread wheat (*T. aestivum*) is an allohexaploid having three genomes (ABD).

Genetic segregation in polyploids differs from diploids in three important aspects. First, an increase in chromosome number permits a corresponding increase in the number of different alleles at any locus. Second, meiotic segregations in a polyploid often lead to sterile gametes, affecting genetic ratios accordingly. In allopolyploids this is further complicated due to three types of pairing in the chromosomes. Supposing there are A_1A_1 A_2A_2 B_1B_1 B_2B_2 D_1D_1 D_2D_2 chromosomes belonging to three genomes, A, B, and D. Three possible type of pairing would be possible:

(i) pairing of identical homologues A_1A_1, B_2B_2, etc., which is called autosyndesis,

(ii) pairing of partially homologues chromosomes A_1B_1, A_2B_2, etc., which is known as allosyndesis,

(iii) pairing of chromosomes belonging to same genome like A_1A_2, B_1B_2, etc., which has no special term. These types of pairing would release tremendous amount of variability never expected in a diploid.

Third, the meiotic segregation in a polyploid, unlike those of diploid, depends on the linkage relationship between gene and centromere and on the frequency with which chromosome associate in multivalent complexes. Summarily, the segregation pattern in polyploids is polysomic and is more complex than diploids. Thus the handling of segregating generations of polyploids has to be different and with much larger F_2 and F_3 populations. Auto and alloloids may be distinguished on the basis of chromosome morphology, chromosome pairing in meiosis, plant morphology and the segregation pattern in the F_2 generation (Table 11.2).

Table 11.2. Criteria to distinguish auto and alloploids

Character	Autopolyploid	Allopolyploid
1. Chromosome number	Multiple of one genome	Multiple of two or more genomes
2 External morphology	Resembling a single diploid ancestor	Resembling two or more ancestors
3. Chromosome pairing	Frequent multivalent formation	Multivalents few due to autosyndesis, mostly univalents
4. Segregation pattern in F_2	Polysomic	Polisomic to disomic

3.1   Effect of Euploidy

Haploids are weak and sterile. Other polyploids have gigantism of plant parts, large cell and stomata but slower growth rate than diploids. Tetraploids are generally vigorous, show polyploid vigour, exhibit more vegetative growth, thicker leaves, bigger flowers and bigger fruits. But they have reduced fertility. Triploids combine the advantages of hybrid vigour and polyploid vigour in most cases. In most species there is a definite level of ploidy, above which added chromosomes or genomes lead to depression of growth and vigour. Presumably, difficulties of nuclear mechanics as well as of other imbalances arising from extreme ploidy level hinder in their establishment in nature. Polyploids have increased flower and fruit size and thus are of interest to ornamental and fruit breeders. Agriculturally important autotetraploids are cultivated potato, coffee, groundnut, berseem, and orchard grasses.

Autotriploids rarely produce functional gametes (pollen or ovule) and the resulting sterility accounts for seedlessness, as in seedless watermelons and lime. All edible bananas are triploids.

3.2   Ploidy in Crop Improvement

Polyploids are abundantly found in the natural state. The fact that more than half of crop plants are polyploids show that polyploidy has played a very important role in speciation. With the discovery that colchicine could induce polyploidy artificially, plant breeders applied this chemical tool in most crop plants to obtain novel and economically superior cultivars. Colchicine stops spindle formation during mitosis and hence one of the two resulting daughter cell has double chromosome number. Once applied to growing tissue, it results in doubling the chromosome number i.e. haploids can be made into diploid and diploid into tetraploid and so on. In general polyploids are gigantic but have slower growth rate, poor fertility and many such associated defects in such artificially induced polyploids.

However, the periods of excitement and despair of inducing polyploids is over and some organised efforts have been made to induce various levels of ploidy in crop plants and to use them in developing superior varieties and saving time in advancing generation. Haploids developed by pollen culture of a F_1 hybrid, if doubled by colchicine treatment, the result into diploid and completely homozygous at all loci. Thus in one generation (just after F_1) completely homozygous plants can be obtained for testing in a hybridisation programme, rather than waiting till F_6 or F_7 generation. This could be great time saver.

3.3   Origin of some Crops Through Polyploidy

Mostly, the species in a genus exist in a series owing to their origin through polyploidy (Table 11.3). But even in one species different chromosome races exist. For example, *Saccharum spontaneum* (wild sugar cane) has somatic chromosome number of 40, 48, 64, 80, 96, 128 and like. The origin of bread wheat (*Triticum aestivum*) is postulated to be as in Figure 11.1. The well known Brassica Triangle (Figure 11.2) is another example of the role played by polyploidy.

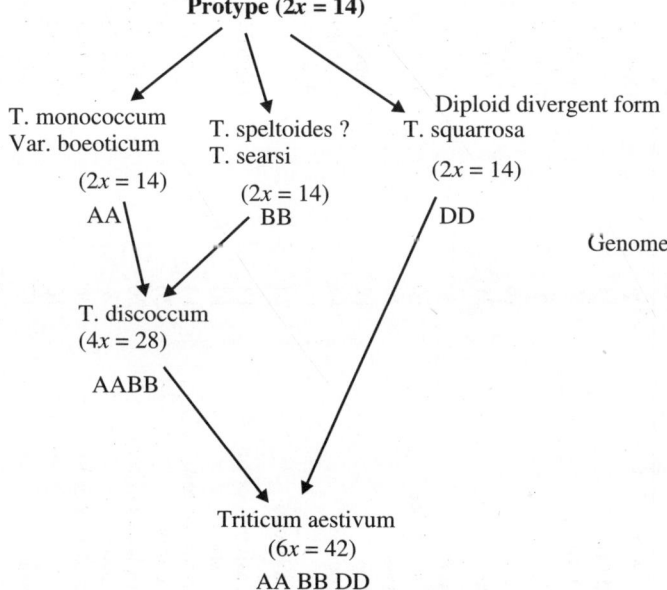

Figure 11.1   Origin of bread wheat (*Triticum of aestivum*) through amalgamation of three distinct species *T. monococcum, T. searsi and T. squarrosa.*

4.   METHOD OF CREATING VARIOUS PLOIDY LEVELS

4.1   Haploids

Pollen and embryosac are the haploid parts of plants. Artificial methods to induce haploids also use these two parts.

Methods using pollen are called androgenic methods or androgenesis and those using ovule are called parthenogenetic methods or parthenogenesis. The unfertilised egg or synergid or antipodals may produce androgenic or androgenetic haploids. One or both of the male gametes contained in the pollen tube may

Table 11.3 Origin of cultivated species from wild ones in some crops through polyploidy.

Genus	Species	Common name	Gametic chromosome (*n*)	Level of ploidy	Genome
Avena	*strigosa*	Sand oats	7	Diploid	A
	barbata	Slender wild oats	14	Tetraploid	AB
	sativa	Cultivated oats	21	Hexaploid	ABD
Triticum	*monococcum*		7	Diploid	A
	dicoccoids		14	Allotetraploid	AB
	aestivum	Bread wheat	21	Allohexaploid	ABD
Sorghum	*versicolor*	Wild sorghum	5	Diploid	A
	vulgare	Cultivated sorghum	10	Tetraploid	B
	halepense	Johnson grass	20	Octoploid	
Nicotiana	*sylvestris*	Wild tobacco	12	Diploid	A
	tomentosa	Wild tobacco	12	Diploid	B
	tabacum	Cultivated tobacco	24	Allotetraploid	AB
Gossypium	*arborium*	Cultivated asiatic cotton	13	Diploid	A_2
	herbaceum	Cultivated asiatic	13	Diploid	A_1
	thurberi	Wild American cotton	13	Diploid	D_1
	barbadense	Sea island cotton	26	Allotetraploid	$A_2D_2D_2$
	hirsutum	American upland cotton	26	Allotetraploid	A_1D_1
Brassica	*nigra*	Black mustard	8	Diploid	B
	oleracea	Cabbage, cauliflower	9	Diploid	C
	campestris	Rapes	10	Diploid	A
	carinata	Gobhi sarson	17	Allotetraploid	BC
	juncea	Rai	18	Allotetraploid	AB
	napus	Rapes	19	Allotetraploid	AC
Saccharum	*officinarum*	Sugarcane	40	Allo-octoploid	

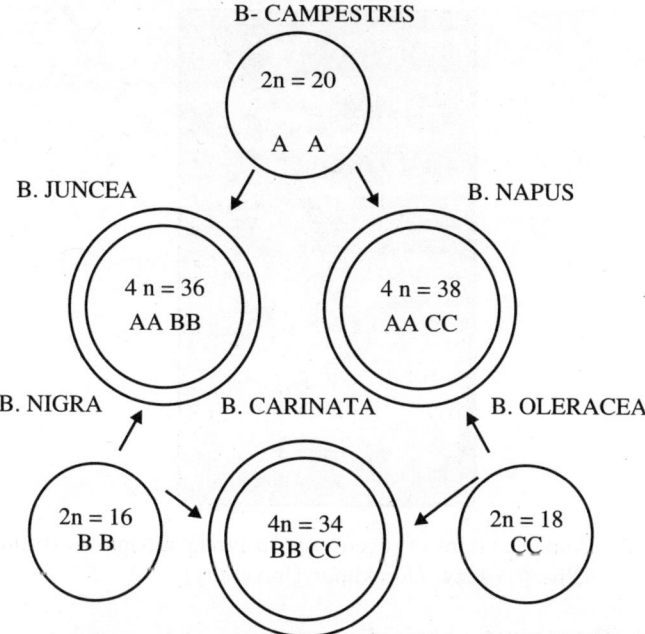

Figure 11.2   Brassica triangle to show the origin of three species of Brassica; namely *B. juncea*, *B. napus* and *B. carinata* from three basic species.

develop by haploid androgenesis after penetrating in ovule. These occur in nature

(a) **Androgenetic haploids:**   Two methods used for experimental production of androgenic haploids, in addition to natural ones are:

   (i) **Pollen and anther culture:**   Pollen or anthers are cultured in artificial media. They form callus and differentiate into haploid seedlings. Anthers may be cultured and the haploid seedlings (Figure 11.3) are obtained. This process may be called 'anther androgenesis' as suggested by Pandey (1973).

   (ii) **Interspecific cross:**   In crosses of *Brassica nigra* × *Brassica campestris* male nucleus of pollen develops into haploid seed after penetrating the nucellus when the egg cell fails. Similar is the case with *Nicotiana tabacum* × *Nicotiana digluta* cross where male gamete develops into a haploid seed. Since the female cytoplasm participates in the development of these haploids the process may be called 'ovule androgenesis' (Pandey 1973).

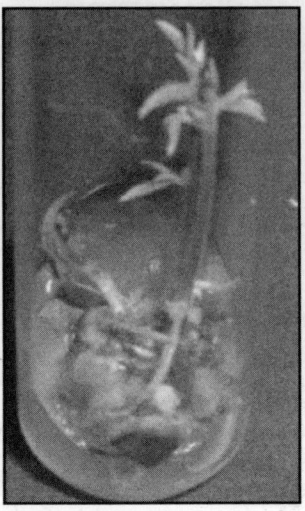

Figure 11.3   Anther culture in pigeon sea to produce haploids (Courtesy: Dr. Dinesh Yadav, Gorakhpur University)

(b) **Parthenogenetic haploids:**

    (i) **Spontaneous (Natural):**   In nature parthenogenesis and haploid polyembryony occur, which are isolated as haploids. Polyembryony is the presence of more than one embryo in a seed. In cucumber, mass screening for haploids is done by floating the freshly harvested seeds on water. The light seeds so separated give a high frequency of haploids, which are believed to have developed parthenogenetically. Polyembryony has been reported in mango, citrus and many other plants. Sometimes only one embryo is diploid and the others are haploids, for example, in rice, cotton.

   (ii) **Temperature shock:**   Rye spikes are exposed to −30C for 30 minutes at 6 hourly interval, after 20 hours of pollination. This results into haploid development of the embryo.

  (iii) **Interspecific crosses:**   Pollination of cultivated barley (*Hordeum vulgare*) with pollen of wild barley (*Hordeum bulbosum*) results into haploid formation. As high as 70 per cent seeds set give haploids. During development after fertilisation, wild barley chromosomes are eliminated and only the haploid set remains. Potato (*Solanum tuberosum*) is pollinated with *Solanum phureja* to obtain haploids. These potato haploids may develop out of the unfertilised egg (haploid parthenogenesis) or antipodal or synergids (haploid apogamy).

(iv) **Mutagen:**   In mutation breeding haploids are obtained. Pollination by pollen treated with N_2O, X-ray and gamma-ray irradiation, toluidine blue etc, produces haploids.

(v) **Specific mutants:**   Crossing with specific mutants, for example, 'stock 6' in maize, results in higher frequency of parthenogenetic development. Colour markers are used such that the diploid grains are red and the haploids are colourless. Chase (1974) has been the pioneer in this technique where a high frequency of haploids is produced.

(vi) **Delayed pollination:**   Delaying the pollination of emasculated buds has been reported to induce the development of unfertilised egg into a haploid.

4.2   Polyploids

(i) **Regeneration method:**   Tetraploid and above levels are induced using the regeneration methods. The growing apex is decapitated and allowed to regenerate. The growing shoot develops callus and new shoot grow out of the callus. Fast cell division takes place in callus formation and many times cell wall formation cannot cope with the rate of nuclear division. If cell wall formation fails between two daughter cells, it gives rise to tetraploid cells. Higher levels of polyploids are also obtained.

(ii) **Temperature shocks:**   Heat and cold shocks given to germinating seeds or developing inflorescence result into polyploids. In both the methods, the principle lies in the failure of formation of cell wall after nuclear divisions. Barley spikes, for example, are exposed to 23-25°C for 20 hours and then to 43°C for 20-30 minutes to get polyploids.

(iii) **Chemicals:**   Colchicine has been very successfully used in solution or paste form on growing points including germinating seeds, young seedling and developing buds to induce polyploids in the past 50 years. Colchicine is an alkaloid extracted from the bulbs of autumn crocus *Colchicum autumnale* and flame lily (Hindi = Kalihari) *Gloriosa superba*. In various plants species, 0.05 to 0.4 per cent concentration of colchicine over a duration of a few hours to days has been used. Colchicine acts through paralysing spindle formation during mitosis, which prevents the migration of the divided chromosomes to the poles in anaphase. In other words, a restitution nucleus is formed

with double number of chromosomes.

Some other chemicals like chloral hydrate, ether, chloroform, etc have also been utilized, though not so successfully.

Alloploids are usually produced by doubling the chromosome number of F1 hybrids between the species. They are also produced by crossing the autotetraploids developed from parental species.

A new man-made cereal named Triticale has been developed by crossing wheat with rye (*Secale cereale*). Octoploid triticale is produced by crossing bread wheat (2n = 42) with rye (2n = 14) and doubling the chromosome of the sterile F_1. Hexaploid triticale is produced in the following manner:

Wheat (*T. durum*; 2n = 28) × Rye (*Secale cereale*; 2n = 14)

$(A_7A_7B_7B_7)$ (R_7R_7)

$$\downarrow$$

F_1

$(A_7B_7R_7)$

Seed does not set in F_1 as the embryo aborts after 15 days. Therefore hybrid embryo is taken out 10 days after pollination and cultured in agar medium using the embryo culture technique. The embryo differentiates into seedlings which are treated with colchicine mutagen to double the chromosome. The doubled seedlings set seed and is called the Triticale. The Triticale has A7A7B7B7 R7R7 constitution, i.e., 14 chromosomes from each of the A, B and R genomes.

5. WAYS TO USE EUPLOID AND ANEUPLOIDS

Depending on the type, these have been used in a number of ways.

5.1 Haploids

Karpechenko (1929) first suggested the use of haploids in plant breeding. They may be utilised as under:

1. By doubling of the chromosomes of haploid, a diploid line homozygous at all gene loci is obtained for one stroke, which thus cuts the time of purification in self pollinated crops. In rice, crosses are made and the anthers of F1 plant are cultured to produce large number of haploids. These are treated with colchicine to produce diploids. All plants thus become homozygous and selection can be done for yield potential and other traits. Thus time required to develop a variety is cut short

by 2/3 years. In barley the F_1 plants are pollinated with pollen from a wild species H. bulbosum. The developing embryo is cultured and seedlings are raised. Most of them are haploids, which are treated with colchicine to produce homozygous diploids. Selection may then be made. Nitsch and Ohyama (1971) for the first time produced a tobacco variety Tam Yah No. 1 by anther culture. This is the best tobacco variety of China today. Thompson (1974) produced Maris Haplona variety of rape by anther culture and doubling the chromosome.

2. Inbreds in maize and other cross pollinated crops can be developed in two years by doubling haploids, whereas in normal course it takes eight years.

3. The juvenile period of fruit trees is large and breeding using any sexual method take long time to breed a variety. But by doubling, the induced haploids would have special significance in cutting short the time.

4. In tetraploids like potato where disease resistance is found in diploid species, haploids of potato can be crossed and the resulting hybrids may be doubled to produce tetraploid and resistant potato.

5. Doubling parthenogenetic monoploid (dihaploids) would give completely homozygous tetraploid potato. This opens avenues of propagating potato by seed, which would be virus free, as virus is not transmitted through seed.

6. Haploids in combinations with mutation are used in the production of entirely new lines (Figure 11.4).

7. Among ornamentals plants, haploids have decorative value due to small flowers and prolonged blooming, the latter being due to sterility.

8. Because there is no dominance in haploids– phenotype is genotype- -they are used successfully in mutation breeding for recovering even recessive mutants in M1 generation.

5.2 Triploids

1. Sterile triploids insure longer flower life and thus a number of varieties have been released in azaleas, lilies, hyacinths and many other flowers.

2. Triploids also provide opportunity to produce seedless fruits and thus in fruits like watermelon, lemon, grape and banana triploid varieties are popular. In fact all edible banana varieties are triploid. Triploid watermelons are seedless, sweet, firm and suitable for transportation. Seeds for triploid watermelons ($3n = 33$) are produced either by hand pollination of tetraploid ($4n = 44$) by diploid ($2n = 22$) male or by

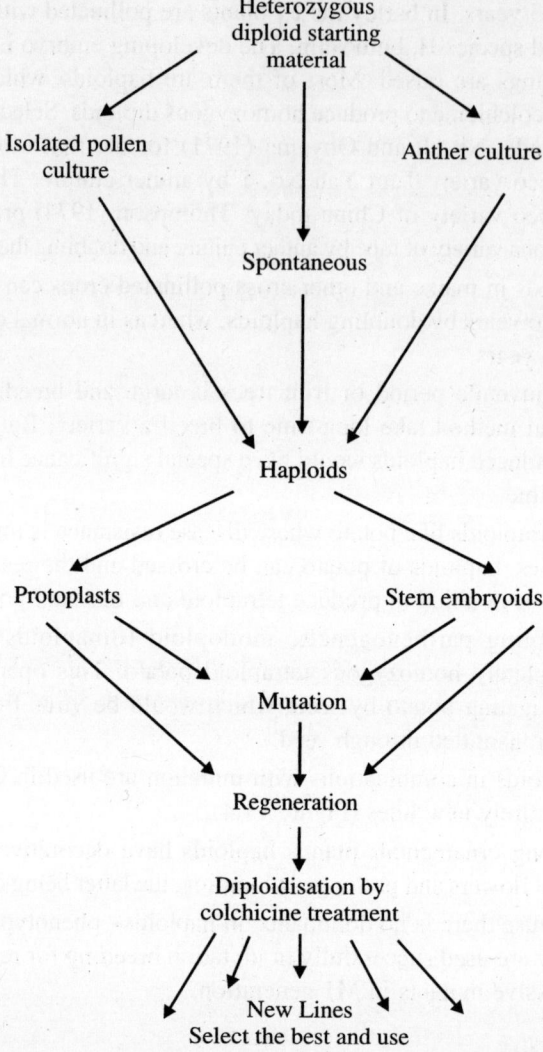

Figure 11.4   Summary of approaches in new line production by using haploids and diploidization by colchicine.

planting tetraploid and diploid in alternate rows. Male flowers of tetraploid plants are pinched out regularly. By cross pollination with diploids, fruits are formed on tetraploids containing triploid seed. On planting, these seeds produce triploid plants which are male sterile. Since fruit setting depends on pollination, in commercial plantation, triploids are interplanted with diploids in the ratio of 5:1. Pollens from diploid plants induce fruit setting in triploids but the fruits remain seedless. A triploid watermelon variety Pusa Bedana was released in 1972 by crossing Tetra-2 (tetraploid from U.S.A.) with Pusa Rasal (a local diploid variety).

3. Triploid hybrids are being developed in sugar beet by crossing cytoplasmic male sterile diploids and tetraploid males. These hybrids combine the advantages of both, hybrid vigour and polyploid vigour. Triploid hybrids yield more sugar than any diploid or tetraploid. In Germany triploid sugarbeets occupy more than 20 per cent acreage. Triploid hybrids are produced by interplanting diploid and tetraploid parents in 1:3 ratio. Seeds thus produced on tetraploid plants are 75 per cent triploid and 25 per cent tetraploid. Male flowers on tetraploid plants in sugar beet are small and hence cannot be removed.

4. Triploid tea, chrysanthemum to produce pyrethrum, poplar and other forest trees are in use as these are better than their diploid counterparts. Pusa Giant Napier grass, which is a triploid involving bajra and Napier grass, is now a widely cultivated fodder variety.

5.3 Tetraploids

1. Increased vegetative luxuriance of tetraploids is utilised in producing fodder varieties like the Pusa Giant variety of berseem. Many varieties of clover, etc., are also in use.
2. Increased flower size has developed a number of tetraploid varieties in Zinnia, Antirrhinum and Petunia.
3. Fruit size becomes also bigger like in tetraploid coffee, grape, groundnut etc.
4. Increased seed size of tetraploid rye has been utilised by forage breeders. Besides, tetraploid varieties have higher protein content and good germination power.
5. In some fruits like grapes, tetraploids have not only bigger fruit size but have less seeds per fruit and well spaced fruits. Many tetraploid varieties of grapes like Pierce are under cultivation.

5.4 Alloploids

1. *Allopolyploids*:   Allopolyploids have given many new species in crops plants like in cabbage family (see Figures 11.1 and 11.2) and wheat on which our civilisation survives. However, in plants of recent origin, the significant example is of triticale which is the first man made cereal by crossing wheat and rye. A number of varieties of triticale

like Armadillo, Badger, UPT 72142 are under cultivation. Triticale is superior to wheat and barley in dry and hilly areas specially. Raphanobrassica was produced by Karpechencko (1927) by crossing radish (2n=18) and cabbage (2n=18) after the chromosome doubling of F1. However this allotetraploid plant proved to be useless as it had roots of cabbage and leaves of radish.

2. *Addition lines*:   The addition of a single alien chromosome (chromosome from a different species) to the full complement of the recipient species is called monosomic addition. Addition of the one pair of chromosome to the recipient species is called disomic addition. Such lines are called monosomic addition line and disomic addition line respectively. For example, in wheat one or more pairs of rye chromosome can be added as illustrated in Figure 11.5. Segregants having one pair of additional chromosomes (wheat+1 pair rye) may be selected as addition line, which breeds true. Alien addition lines of wheat having III rye chromosome are immune to powdery mildew. But as such lines have no place as a variety due to poor stability, reduced fertility and undesirable feature. However, alien additions of small segments have been a success.

An alien-addition line carries one chromosome pair from a different species in addition to the normal somatic chromosome complement of the parent species. Generally, the purpose of alien-addition is the transfer of disease resistance from related wild species, e.g., transfer of mosaic resistance from Nicotiana glutinosa to *N. tabacum*.

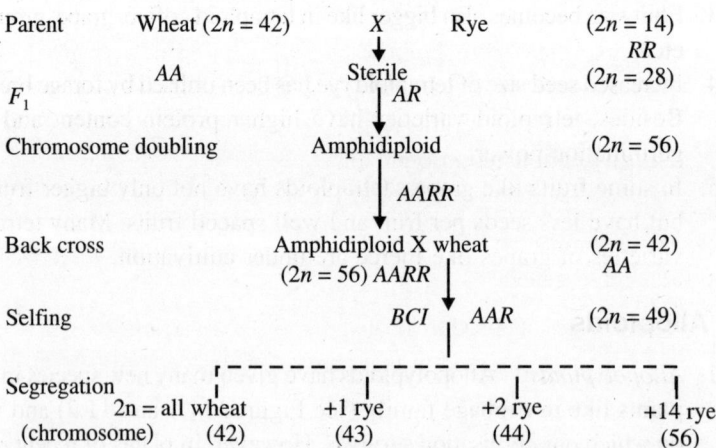

Figure 11.5 Additions of one or more pairs of rye chromosome in wheat

Substitution lines: Substitution lines are those with one or more pairs of substituted chromosome by another species or variety. For example, one or more pairs of wheat chromosomes may be substituted by rye chromosomes such that substituted lines have 2n=42 (40 wheat +2 rye and so on) chromosomes. Chromosome number 5 of rye (5R) substitutes wheat chromosome 4A very well. A variety, 'Blaukorn' of West Germany, has a full 5R chromosome in place of 4A wheat chromosome. This variety is rust resistant and has hairy neck (Zeller and Baier, 1973). A wheat substitution line carrying one chromosome of Agropyron has been released as a variety in Germany by the name Weique (Wienhues 1965). Alien substitution line has one chromosome pair from a different species in place of the chromosome pair of the recipient species. Mosaic resistance gene was transferred from *N. glutinosa* to *N. tabacum*; the resistant *N. tabacum* line had 23 pair of *N. tabacum* chromosome and one pair (chromosome H) of *N. glutinosa chromosomes*. Both alien- substitution as well as addition are effectively viable in polyploid species only.

5.5 Aneuploids

Aneuploids have been extensively used in genetic studies basically on the ground that deficient or extra chromosomes give aberrant segregation ratios. A detailed account of the use of aneuploids has been given by Khush (1973). These are summarised below:

1. Monosomics have been used to produce substitution lines, as described on page 184.

2. Monosomics have been used to locate genes on a particular chromosome. Nyquist (1957) located genes for rust resistance on chromosome XIII by this technique. In F_2 of crosses between resistant and susceptible monosomic lines only the monosomic would not give expected segregation ratio on which the particular gene is located.

3. Monosomics have been used to transfer the chromosomes carrying a desirable gene.

4. Nullisomics are used to locate genes on chromosomes as the character will not be expressed in that particular nullisomic plant.

5. Nullisomics are used for chromosome substitution.

6. *Primary trisomics* (lines with one extra chromosome of the normal chromosome complement) have been used to assign genes to a specific

chromosome (Figure 11.6). Crosses are made with "aa" to all primary trisomics of that species and segregation pattern in F_2 is observed. Typical 3A : 1a ratio would be observed in all the crosses except in the critical cross i.e. the cross involving the trisomic on which "a" gene is located.

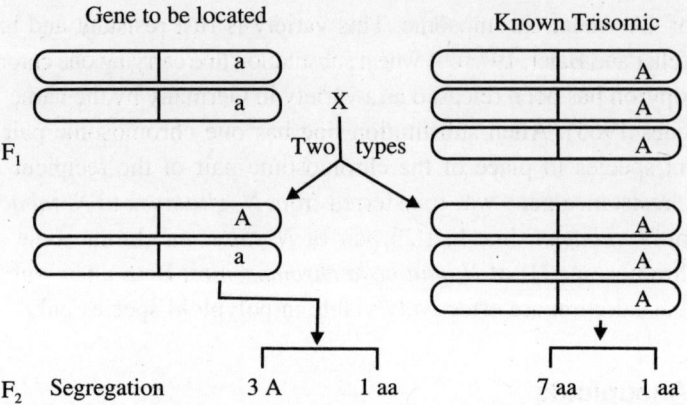

Overall less aa and disturbed segregation ratio.

Figure 11.6   Segregation in a critical cross of disomic and trisomic lines

7. *Secondary trisomics* are the one where the extra chromosome has identical arm i.e. isochromosome. These are used in genetic studies.
8. *Tertiary trisomics* are the one where the extra chromosome is constituted of two chromosomes. Balanced tertiary trisomics are used for the production of hybrid seed in barley.

6. USEFUL REFERENCES

Allard, R. W. 1960. *Principles of Plant Breeding*. New York; John Wiley and Sons Inc. 485 pp.

Chopra, V. L. (Ed) 2001. *Breeding Field Crops*. Oxford & IBH Publ. Co. New Delhi, Calcutta.

Kasha, K. J. 1974. *Haploids in Higher Plants*. Canada; University of Guelph, 421 pp.

Khush, G. S. 1973. *Cytogenetics of Aneuploids*. New York: Academic Press, 301 pp.

Maheshwari, P. 1950. *An Introduction to the Embryology of Angiosperms*. London; McGrawHill, 453 pp.

Simonds, N. W. 1979. *Principles of Crop Improvement*. Longmans, London and New York.

Khokholov, S. S. 1976. *Apomixis and Breeding*. New Delhi, Amerind Publ. Co. 346 pp.

Stebbins, G. L. 1949. Types of polyploids: Their classification and significance. Adv. Genetics 1: 403–409.

Breeding for Disease and Insect Resistance

1. DEFINITION AND HISTORY

Resistance is the capacity of plants to resist, withstand lessen and overcome the attacks of parasites. Without resistance to parasites, plants would have not been able to survive. Parasites are organisms which derive their food from the plants on which they attack. Some parasites, in addition to causing physical damage, create diseases and thus are called pathogens.

Annual crop losses were estimated in the year 2007 amounting to Rs, 1,40,000 crores by ASOCHAM due to pests and diseases. India falling in the tropical zone has to face more pest and disease problems as would be clear from Table 12.1.

Agriculture is the most unnatural phenomenon on earth's surface, which disturb the age-old balance of the host-parasite system. Human being try to protect the crops by using resistant varieties or agro-chemicals, and parasites evolve to overcome it. This fight goes on and sometimes results into epidemics and heavy crop losses.

Parasites and their host plants must have originated and evolved together. The Greek philosopher (3rd century B.C.) was the first to record varietal

Table 12.1 Number of pests and diseases reported in tropics and temperate zones (Swaminathan, 1979)

Crop	Temperate zone	Tropical zone
1. Rice	54	500 – 600
2. Maize	85	125
3. Citrus	50	248
4. Tomato	32	278
5. Beans	52	250 - 280

differences for diseases reaction. For quite some time diseases were thought to be the result of rotting of plant parts rather than the cause of rotting. Damage by insect was not doubted as one could see them eating the plant parts. Prevost (1807) reported that bunt disease of wheat is caused by a fungus, and De Bary (1861) established the parasitism of stem rust on wheat. Biffen (1905) was the first to report that the resistance to yellow rust of wheat is governed a by recessive gene, as by the time Mendelian laws were rediscovered already. He reported a 3:1 ratio in F_2 for susceptibility to resistance. Systematic resistance breeding began only after that. Still scepticism continued over resistance breeding due to "Bridging-host" hypothesis of Ward (1902). This hypothesis assumes that pathogen is plastic and it can be influenced by the host substrate to change its pathogenicity. However, brave observation of Biffen (1912) and others who demonstrated the stability of the host resistance, sustained resistance breeding programmes. Today resistant varieties are the foremost contribution of plant breeding to stabilise the production and contain the parasites. In fact stability of production depends on the resistant varieties.

2. DIFFERENCES BETWEEN PRODUCTIVE AND RESISTANCE BREEDING

Productive breeding, the genotype of the plant and its interaction with the environments are the two considerations (G × E); while in resistance breeding there are three components (G × E × P), known as the "disease triangle" (Figure 12.1). Interaction of these three components produces resistant or susceptible reaction. Since the triangle is equilateral, all the three components have equal importance. Thus resistance is observed when either weather is unfavourable or pathogen is avirulent or host has resistant genes. Conversely, virulent pathogen, favourable environment and susceptible host are essential for disease to occur. An avirulent race of a pathogen is the one which can not cause the disease, and the one which can is called virulent. A pathogen race or pathotype is said to be virulent if it is able to attack a host with specific resistance gene, and *avirulent* when it can't attack. *Pathogenicity* is the ability of a pathogen to attack a host and is synonymous to *virulence*.

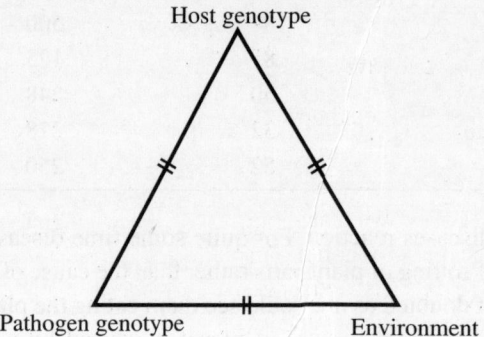

Figure 12.1   The *disease triangle* with genotypes of host and pathogen, and environment of equilateral arms demonstrating equal importance in disease development.

2.1 Productive breeding is concerned with the increase in yield potential per se, while resistance breeding it to provide stability of yield. Thus a resistant variety yields the same in presence or absence of high disease pressure.

2.2 Artificial epiphytotics i.e. artificial creation of high disease pressure are necessary to effect selection in resistance breeding while these are not needed in productive breeding programmes.

2.3 Yield potential once incorporated in a genotype becomes permanent feature while resistance reaction may be lost mainly due to development of virulent races.

3.   LOSS OF RESISTANCE

Resistant varieties may become susceptible after some time, which is called break-down of resistance. This may be due to any of the following reasons:

3.1 Change of Environment

Resistance genes function under a given set and range of environment. Once this range is crossed, resistant genes do not function to provide resistance and the resistant variety becomes susceptible. For example tomato varieties carrying Cf1 gene show resistance to *Cladosporium fulvum* fungal disease in summer but not in winter.

3.2 Resistance erosion

The host may lose the resistance gene either due to mutation, out-crossing with a susceptible variety or segregation. This type of loss is rare though and is known as resistance erosion.

3.3 Appearance of new races or biotypes

The disease causing pathogen may create new races or insect-pests may create new biotypes. The resistant variety thus may become susceptible. This type of loss of resistance is very common. For example, IR26 variety of rice was resistant to biotype1 of brown plant hopper pest. Then a new biotype 2 appeared to which IR26 became susceptible and in 1974 farmers in Kerala suffered heavy losses. There are clear evidences of how races of brown rust of wheat have changed with the change of varieties. Races 162 and 162A were predominant around 1965 but changed in 1975; races 12, 17 and 104 became predominant. Wheat varieties Kalyansona and Sonalika were released in 1967 and were completely resistant but gradually newer races like 12, 77, 104 appeared to which these varieties became susceptible (Table 12.2). Janak was released in 1973, thereafter race 10 reappeared to which Janak became susceptible. A new variety Girija was released in which was resistant all leaf, stem and yellow rusts but became susceptible within 10 years. One of the most popular variety PBW343 released in 1995 thereafter also fell susceptible in 2001 and had to be replaced by HD2733. After a span of 10 years HD2733 has also to be replaced in North West Plain Zone due to appearance of a new pathotype (race) by HD2967 in 2011. Let us see for how long before a new race appears. Thus the game of disease and breeders goes on.

Sometimes races also create doubt about the resistance as would be clear from Table 12.3. Bean varieties, White Marrow and Robust, are resistant to Alpha and Beta strains when tested singly. But when a mixture of Alpha and Beta races are used, both the bean varieties become susceptible. In nature across locations, the pathogen occur as racial mixture of various composition. Change of place, therefore, changes disease reaction i.e. resistance.

4. PLANT BREEDER VERSUS NATURE

By now it is clear that plant breeders breed new resistant varieties and pathogen make these susceptible by creating new races or biotypes. A plant variety remains resistant only as long as there are no races or biotypes (virulent ones) to attack it. Apparently new races or biotypes are created by nature in response to plant breeders' resistant varieties. Thus it seems that the two are responding

Table 12.2 Reaction of some wheat varieties to races of brown rust (Reddy and Rao, 1977) S = susceptible, R = Resistant

Race Variety (gene)	10	11	12	17	20	63	77	104	106	107	108	162	162A
Kalyan Sona (Lr113, Lr14B, Lr18)	S	R	S	S	S	R	S	S	R	R	S	S	S
Sonalika (Lr29, Lr11, Lr13)	R	R	R	R	R	R	S	S	R	R	S	S	S
Janak (Lr3, Lr15)	S	R	R	R	R	R	R	R	R	R	R	R	R
Girija (Lr10, Lr15)	R	R	R	R	R	R	R	R	R	R	R	R	R

Table 12.3 Dependence of disease resistance on races (Mc Rostie, 1921)

Variety	Alpha strain	Beta strain	Mixture of Alpha & Beta strains
White Marrow	Resistant	Susceptible	Susceptible
Robust	Susceptible	Resistant	Susceptible

to each other like in a war game. Let us see their relative strength. Table 12.4 lists the relative superiority of pathogens over host on types of mechanism to create genetic variability.

In addition to the haploid and diploid nuclear cycles of the host, the pathogens have dicaryotic cycle where two nuclei occupy the same cell without fusion. The host range of a dicaryon is different than a haploid or diploid. Similarly, in addition to a sexual and asexual mode of reproduction and rare parthenogenetic, the pathogens have five additional life cycles. Parthenogenetic mode of reproduction is found in some pests. All these modes of life cycles and additional modes of creation of variability give more plasticity to the pathogens and give tremendous power over their hosts to create new races and create them.

5. RACE AND ITS IDENTIFICATION

Race of a pathogen is a mere combination of virulent and avirulent genes, which are identified on a set of differentials or host testers. Stakman (1914) and his group are the pioneers in the concept of races and have reported their existence based on the reaction of pathogens on a set of differentials, also called host testers. Races are physiologically specialised groups in a *forma speciale*. For example, the fungus causing black rust or stem rust of wheat is classified as:

Genus:	*Puccinia*
Species:	*Puccinia graminis*
Forma speciale:	*Puccinia graminis* f. sp. *tritici*
Race:	*Puccinia graminis* f. sp. *tritici race* 15-1

Wheat scientists in India are using the term Pathotype in place of Race currently. It is based on Binomial System of nomenclature like old **race 15-1** of wheat rust is now designated as **pathotype 123G15**. Greater the number more virulent is the pathotype. *A differential variety or a host tester is a host with known major gene, used specifically for differentiating races or pathotypes.* There is a minor difference in race and pathotype. The differentials in case of race identification may not have known genes but in case of pathotypes they have indentified resistance genes. Differentials with only gene for resistance (monogenic) are most efficient to identify pathotypes. A series of such hosts with known single gene for resistance are used for differentiating pathotypes. If there are n number of differentials and 2 levels of disease reaction (resistant and susceptible) then 2^n races can be identified. For example, with 4 differentials and 2 level of disease reaction (resistant and susceptible), a total of $2^4 = 16$ races can be identified. Consider the reaction of races 10, 11 and 12 on the

Table 12.4 Relative ability of host and pathogens for creating new genotypes

Ground		Host		Pathogen
Nuclear cycle	1.	Haploid (n)	1.	Haploid (n)
	2.	Diploid ($2n$)	2.	Diploid ($2n$)
			3.	Dicaryoitic ($n + n$)
Life cycle	1.	Asexual	1.	Asexual
	2.	Sexual	2.	Sexual
	3.	Parthogenetic	3.	Parthenogenetic
			4.	Haploid – restricted dicaryotic
			5.	Haploid – dicaryotic
			6.	Dicaryotic
			7.	Haploid – Diploid
			8.	Diploid
Creation of genetic variability	1.	Mutation	1.	Mutation
	2.	Hybridisation and Recombination	2.	Hybridisation and Recombination
	3.	Genetic engineering	3.	Heterokaryosis
			4.	Parasexual cycle
			5.	Cytoplasmic adaptation
			6.	Heterocytosome

given 4 varieties if wheat (Table 12.2). Race 12 can attack only Kalayan Sona, race 10 can attack Janak also, while race 11 can attack none. Thus these races can be identified based on their reaction on the differentials. In India a set of 25, 22 and 21 differentials are standardised for leaf rust, stem rust and stripe rust respectively (Singh et al, 2011). Based on 12 differentials Stakman reported 300 races of *Puccinia graminis* f. sp. *tritici* on wheat.

6.   CLASSIFICATION OF RESISTANCE

Each plant species is attacked by hundreds of kinds of pests and pathogen (Table 12.1). Even a single plant is attacked by hundreds or thousands of individuals of a pathogen. Plants express their reaction and manage to survive using various mechanisms that can be classified variously.

A.   Based on Existence

(1) *Preformed*

When the resistance is already present in the plant even in the absence of the pathogen, it is known as preformed, or axenic or passive type of resistance. For example powdery mildew resistance in barely variety Nigrete or rust resistance in some wheat varieties exists even in the absence of the pathogen.

(2) *Induced*

When the resistance is not present in the host plant in absence of the pathogen, it is called induced resistance or active resistance or apergic resistance. For example, cotton seedlings are susceptible to many pests due to absence of gossypol but once infected by Verticillium, seedlings synthesise gossypol by induction and become resistant. Normally only mature cotton plant has gossypol.

B.   Based on Type of Host Response

(1) *Immune*

Immune may be defined as "exempt from infection" or 100 percent freedom from disease. No symptom develops in case of immune response. For example, potato is immune to wheat rust.

(2) *Tolerance*

Tolerance may be defined as the inherent or acquired capacity of the host to endure disease (Nelson, 1973). Tolerance may be defined either as symptomless carrier in viral diseases where the virus multiplies in the host tissue but the disease symptoms do not appear. Tolerance also means the disease tolerance where yield losses are minimal even under maximum presence of the disease. Therefore, tolerance is the capacity of the host genotype to compensate yield losses even when diseased. There can't be tolerance against the disease that affects the end product like bunts etc. This is due to the reason that the host gets no chance to recover the loss by way of compensation (Schafer, 1971).

(3) *Resistance*

In strict sense resistance characterises those situations in which some degree of host - pathogen interaction is evoked. Therefore, resistance is always partial and some symptoms do appear on the host, contrary to immunity, where no symptoms appear. This is the reason that resistance is always "scored" on a scale 0 to 9 or 1 - 10, depending on the crop and disease.

C. Based on Genetic Control

(1) *Oligogenic*

Oligogenic (Oligo = a few) also known as major genic resistance is governed by a few genes whose individual effect is very clear. For example, the classical case of yellow rust resistance in wheat reported by Biffen (1905), R 1 to R 11 genes in potato blight or Xa genes against bacterial blight disease in rice etc. Resistance may be governed by dominant or recessive genes.

(2) *Polygenic*

Polygenic (poly = many) also known as minor genic resistance is governed by many genes whose individual effects are not marked as contrasted in case of oligogenic. Polygenic resistance is like quantitative character. For example Lutpon and Johnson (1970) reported that in wheat, cross of a susceptible variety with a resistant variety Little Joss, F_1 was susceptible. In F_2 generation a normal curve of segregants was obtained. Thus it was concluded that yellow rust resistance of Little Joss is polygenic.

(3) *Cytoplasmic*

Resistance of this type is governed by cytoplasmic factors. For example, all Tms (Texas male sterile) cytoplasm of maize used in hybrid seed production is susceptible to T race of Southern Corn blight in USA. Irrespective of nuclear genes, any maize hybrid behaves as susceptible to T race, which was the reason of major epidemy in USA. Contrary to Tms, C or S cytoplasm are also male sterile but resistant to T race of the pathogen.

D.   Based on Mechanism of Resistance:

(1) *Mechanical*

Mechanical or structural resistances are rendered due to the external or internal peculiarities of the host plant. The first line of defence against the pathogen is the surface. Cuticle, wax and thick hairs on plant parts and foliage check the germination and entry of the pathogen in the host tissue. The epidermal cells of the rice varieties resistant to blast fungus are lignified. Formation of cork layer beyond the infection point of *Rhizactonia solani* in potato, or abscission layer around the diseased spot in peach against *Xanthomonas pruni* are a few more examples to check the further spread of the pathogen beyond the infection point.

(2) *Biochemical*

A series of defence reactions are evoked in the host following invasion by the pathogen. Some resistant reactions are as follows:

(i) **Hypersensitive:**   Ward (1902) first focused attention on hypersensitivity as a potential defence reaction. It is due to the increased sensitivity of the host cells in the vicinity of the infection site and is expressed as necrotic spots. The necrotic tissue separates the normal living tissue of the host plant from the parasite and thereby checks its further growth. Muller (1959) described the hypersensitivity as "*all morphological and histological changes that when produced by an infectious agent, elicit the premature dying off (necrosis) of the infected tissue as well as inactivation and localization of the infectious agent*". Resistance of potato to late blight disease caused by fungus *Phytophthora infestans* is of hypersensitive type.

(ii) **Resistance to toxins:**   Some pathogens produce toxins, which kill the host or host tissue. Resistant hosts in such cases are non-sensitive to toxins. For example, causal organism Victoria Blight called *Helminthosporium victorae*, produces a toxin named Victorin. Victorin can kill the oat plant. Resistant varieties are resistant to the toxin.

(iii) **Phytoalexins:**   *Phytoalexins may be defined as substances formed by the host plant in response to injury, physiological stimuli, infectious agents or their products that accumulate to levels which inhibit the growth of microorganism and repel insects and nematode* (Day, 1974). For example *DIMBOA* in onion and maize, *pisatin* in pea, *phaseollin* in beans, *rishitin* and *phytotuberin* in potato are the phytoalexins and provide resistance.

E.   Based on Degree and Range of Resistance:

Van der Plank (1963, 1968) introduced these concepts and classified resistances in two groups; vertical and horizontal.

(1) *Vertical Resistance*

Vertical resistance is also known as differential resistance, or race specific resistance. It is represented by the vertical Y axis of once disease reaction is plotted in a histogram (Figure 12.2). Resistance gene functions against some races while against others it does not function.

(2) *Horizontal Resistance*

Horizontal resistance is also known as uniform resistance or race non-specific resistance. It corresponds to horizontal line of the X axis once the disease reaction is plotted in a histogram (Figure 12.3). This type of resistance provides resistance against all the prevalent races of a pathogen.

7.   GENETICS OF HOST - PATHOGEN INTERACTION

The first successful attempt to understand the genetics of host - pathogen interaction was made by Flor. He published his conclusion in 1956 in the form of a "gene-for-gene" hypothesis. This hypothesis states that "for each gene conditioning the resistant reaction in the host, there is a specific gene conditioning pathogenicity in the pathogen". Flor worked with linseed (flax) and its rust disease caused by *Melampsora lini* system. In this system resistance is dominant to susceptibility and avirulence is dominant to virulence. That means that resistance in linseed and avirulence in the rust is governed by dominant genes. In this situation resistant reaction is observed only when the host with a dominant gene is attacked by a pathogen with a dominant gene. In other combinations, it will be a susceptible reaction (Table 12.5).

Table 12.5   Disease reaction in linseed (flax) and linseed rust system

Pathogen	Host	Linseed genes for resistance	
		Dominant	Recessive
Rust genes for	**Avirulent (Dominant)**	Resistant	Susceptible
causing disease	**Virulent (Recessive)**	Susceptible	Susceptible

Day (1974) reported that in 27 host - pathogen systems, the gene-for-gene system functions. It has also been reported in wheat and hessian fly system also. Hessian fly is a problem pest of wheat in USA.

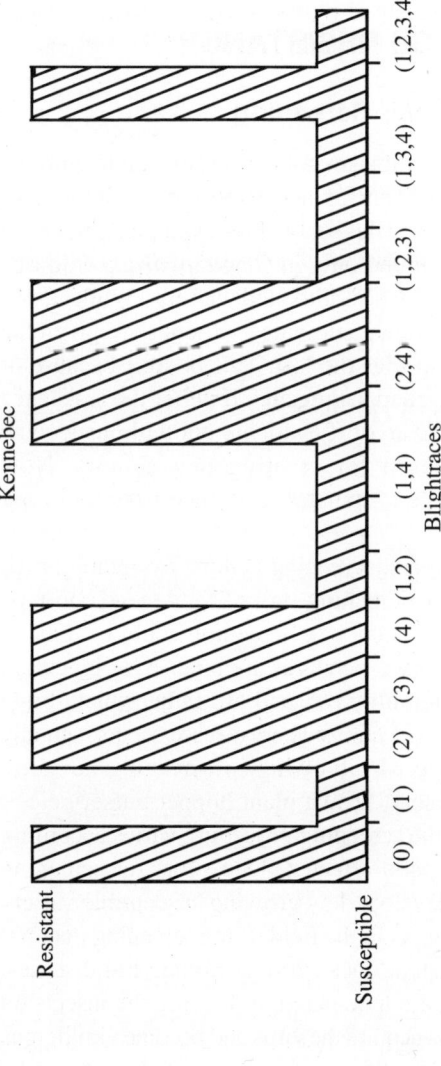

Figure 12.2 Vertical resistance against races of *Phytophthora infestans* in potato variety Kennebec.

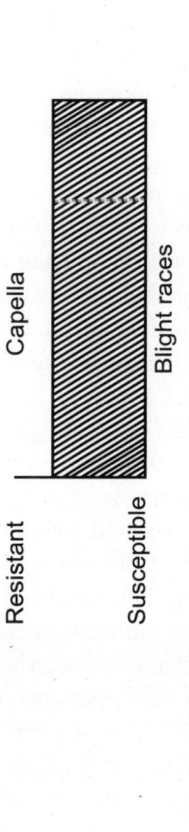

Figure 12.3. Horizontal resistance against races of *Phytophthora infestans* in potato variety Capella as expressed in uniform and moderate level of resistance.

Van der Plank introduced the term *Vertifolia Effect*, which refers to epidemic development in a variety carrying vertical resistance gene. The term is derived from the potato variety Vertifolia which suffered heavy losses in Europe due to potato blight. Probably the level of horizontal resistance was low and the blight fungus had developed virulence against the vertical resistance.

8.   BREEDING FOR DISEASE RESISTANCE

8.1   Resistance Donors and their Screening

Resistance donors may be found in germplasm collections available with the breeder or national and international agencies and gene banks. Sometimes resistance donors are found in wild species only. For example, resistance against grassy stunt virus of rice was found only in *Oryza nivara*, a wild rice found in eastern Uttar Pradesh, and in no cultivated variety. While using resistance donors in a breeding programme, cultivated varieties are given the first choice, and wild and unrelated species the last. This is done because of ease in hybridisation and recovery of good segregants. If either the resistance is not available in the existing germplasm or if available but with inadequate level, then mutation breeding is used for creating new donors. Now biotechnological methods are available to transfer resistance from unrelated species and genera.

In order to identify resistance donors, screening is done by creating high disease pressure i.e. epiphytotics. The screening procedure depends on type of disease, nature of crop, place of screening etc. Screening could be done in laboratory or field condition. Laboratory procedures are precise for inoculum load, control of light, temperature, humidity etc. and are rather quick. Field screening is done usually on "hot spots" (locations where disease occurs naturally every year with considerable intensity) by growing in screening nurseries like rust nursery, bacterial blight nursery, brown plant hopper nursery etc.

Table 12.6 gives an idea of how the screening procedure varies according to disease and host plant and plant parts affected. "Sickplots for screening wilt and root attacking diseases are developed by growing susceptible variety year after year and even adding inoculums in the field. If the screening material is grown in such plots, only the resistant ones survive. Some viral diseases, which spread through insect vectors, are inoculated by feeding the insects on diseased plants. By feeding, such insects acquire the virus and become viruliferous i.e. capable of spreading the virus. Viruliferous insects are released on the material to be screened. Thus the screening procedure is varied for soil-borne, air-borne, seed-borne and vector-transmitted diseases.

8.2   Incorporation of Resistance

The methods used for developing resistant varieties are almost the same as for productive breeding. Some methods are described below, indicating specific points where they differ from normal breeding. If the resistance is governed by major genes, it involves a different handling of the segregating populations than if it was governed either by recessive or polygenes. In case of recessive genes, proportion of resistant segregants will be very low. In case of polygenes, probability of recovering homozygous segregants will be extremely low and thus selection in F_2 should be avoided. Any disease pressure will destroy the heterozygous segregants.

8.3   Introduction and selection

Resistant donors or varieties are introduced from different locations in the country or abroad. *Pusa Sawani* variety of lady's finger (Okra) which resulted out of an introduction, has been the only variety resistant Yellow Mosaic for decades.

8.4   Hybridisation and Selection

This method is to combine the resistance of one or more donors (D) into an acceptable agronomic type (A). Two way $(D_1 \times A) \times D_2$ or multiple $(D_1 \times A)$ $\times (D_2 \times A)$; $(D_1 \times D_2) \times (D_3 \times A)$ crosses are made to effect selection in the segregating generations. Either the pedigree, bulk (single seed descent) or back cross method is followed.

   (i) **Pedigree:** In the pedigree method resistant plants are selected from the F_2 population after creating epiphytotics. Single plant progenies of selected plants are screened for a few generations. Epiphytotics are created in F_2 itself if resistance is dominant, otherwise in F_3. The back cross was first proposed by Briggs (1930) as a means of incorporating disease resistance by crossing of the recurren parent repeatedly with the hybrid progeny for the purpose of recovering its characters, except the addition of disease resistance from the non-recurrent parent.

   (ii) **Back cross:** The back cross technique is very simple in the case where disease resistance of the non-recurrent parent (donor) is of high degree and governed by one dominant gene. If the recurrent parent has acceptable agronomic base and the crossing is easy, the technique becomes much simpler as outlined in Figure 6.3. In each back cross generation, crossing is made with resistant (Rr) plants. Rr plants are identified from rr plants by artificial inoculation with pathogen. Six to eight back crosses are made before the resistant (Rr) plants are selfed to get resistant (RR) plants with other characters of the recurrent parent.

Table 12.6. Crops, their major disease, method of screening for t and resistance donors

Crop Disease	Type of disease	Method of screening	Donors
Rice:			
1. Blast	Fungal-foliage	Spray spore suspension	Tadukan, Tetep
2. Leaf blight	Bacterial-foliage	Spray suspension, pin-prick and clip	BJ1, LZN, UPRB 30
			O. longistaminata
3. Tungro	Viral	Release viruliferous vectors	*O. nivara*
Wheat:			
1. Rusts	Fungal-foliage *Aegilops*	Dust or inject spores	*T. teempopheevi*,
2. Loose smut	Fungal-seed	Inoculate florets	--
Maize:			
1. Stalk-rot	Fungal-bacterial-rot	Inject spores in 2nd internode before dawn	--
2. Downy mildew	Fungal-foliage	Put piece of infected leaf in whorl	--
Cotton:			
Wilts	Fungal-Nematode soil borne	Grow in 'sick-plots'	*G. barbadense*, Hartsville
Tobacco:			
Mosaic	Viral	Rub infected leaf extract on healthy leaf	*N. glutinosa*
Oats:			
Victoria blight	Fungal-foliage	Germinate seed in fungal toxin	Landhafer, Bond
Sugar cane:			
Smut	Fungal-foliage	Dip stem cuttings in spore suspension and then plant	Co 449 Co 527
Tomato:			
Wilt	Fungal-soil borne	Dip seedling roots in inoculum and then transplant	*L. pimpinellifolium*
Root knot	Nematode-soil borne	Inoculate seedlings grown in test tube with egg masses	----do----

If the resistance of the non-recurrent parent is governed by recessive genes (Figure 6.4), slight modification is done. The progeny from each back cross would segregate into two genotypes (RR and Rr) both of which would be susceptible and identical. The difficulty is over come by selfing either simultaneously or in alternate succession of back crosses. In the former case, a number of plants (both RR and Rr) are back crossed to the recurrent parent and selfed at the same time. The selfed progenies are tested if these segregate for disease resistance. Back cross progenies of all those plants which segregate are kept, rejecting non-segregating ones. In the latter case selfing is done once and then tested for segregation for resistance. Only segregating progenies are back crossed. This procedure takes more time though the principle in both is the same. In cross pollinated crops, the heterozygous back cross hybrids are crossed among themselves and homozygous lines are recognised by the fact that they do not segregate in crosses with the double cross. Orton (1907) was probably the first to use back cross to develop the wilt resistant variety of watermelon named conqueror.

(iii) Composite cross: In the technique breeding for multiple disease resistance is aimed at and F_2 seed of many crosses involving a number of donors is grown together. The material is handled by the composite method and disease epiphytotics are created. For example, composite cross CCXXV of barley developed in 1967 by California University was aimed to incorporate resistance to yellow dwarf and mildew.

(iv) Recurrent selection: Jenkins et al. (1954) proposed recurrent selection as a method for concentrating genes for resistance to *Helminthosporium turcicum* leaf blight of maize. This procedure can also be followed in self pollinated crops with the help of male sterile gene. Male sterility would promote out crossing and just by inoculating a disease in the field, resistant genes could be concentrated. Susceptible plants may be uprooted before flowering.

(v) Bulk method: If the resistance is polygenic, or epiphytotics failed in a particular generation, material may be advanced either by bulk sample or by single seed descent method.

8.5 Mutation Breeding

(i) **Defect elimination:** When a well adapted variety becomes susceptible, the quickest procedure may be to induce a resistant mutant in it.

(ii) **De novo source:** If the resistant donor is not available in the crop or related species, mutation is the only source. For example, ergot resistance in bajra is not available and mutation is the only source to create resistance.

(iii) **Breaking undesirable linkage:**   If resistance has bad linkage either with susceptibility for another disease or undesirable character, the mutation approach may be utilised. For example, the resistance to crown rust of oat governed by PC-2 gene is closely linked to HV gene responsible for susceptibility to victoria blight. Mutation was induced to develop varieties resistant to both diseases. Varieties resistant crown rust were treated with mutagen. The M_2 seed was screened with the help of victorin toxin produced by the victoria blight pathogen (*Helminthosporium victoriae*). M_2 seeds are grown in a moistened paper towel. When germination initiates, the seeds are sprayed with victorin toxin and allowed to germinate for 3-4 days. The toxin kills the roots of susceptible seedlings but the resistant ones are not affected. The resistant ones are transplanted and seeds are obtained. The frequency of the resistant ones is one in 1000 M_2 seeds. Thus varieties were developed with combined resistance to crown rust and victoria blight.

(iv) **Transfer from alien sources:**   Transfer of resistance from alien species by induced translocation: Sears (1956) developed a procedure where rust resistance from Aegilops was transferred to wheat by irradiation-induced translocation of Aegilops chromosome.

Using mutation breeding, 13 varieties of cereals, 9 of legumes and 2 of fruits and flowers, have been released which are resistant to a number of pests and diseases (Sigurbjorasson and Micke, 1974).

9.   BREEDING FOR INSECT RESISTANCE

More than half of the species of insects derive their food from living plants, and in that effort they inflict severe damage to agricultural crops. The 1974 epidemy of brown plant hopper on rice in Kerala and many States are example of the extent to which crops can be damaged by insects. There was no period of confusion in identifying damage due to insects, contrary to diseases, as one could see an insect eating the plant parts. Major differences between disease and insect resistance are given in Table 12.7.

9.1   Mechanism of Insect Resistance

There are three mechanisms (Painter 1952) by which hosts resistance to insect-pests operate, one or more of which may be frequently resent in resistant varieties.

(i) **Non-preference:**   The preference and non-preference operates through two modes, one through the host finding mechanism and other through taste. Resistant variety may lack attractive stimuli or may possess repellents. Preference or non-preference is due to a number of factors, which are inherited. Some variety which insects do not prefer to eat naturally becomes resistant.

Table 12.7   Difference in disease and insect resistance

Diseases	Insect-pests
1. Organism has simple organisation and is small	1. Complex and large
2. Host finding mechanism is based on chance	2. In addition, senses play important role in finding host
3. Movement of pathogen at the end of the life cycle only	3. Frequent movements by insects
4. Host specificity more marked	4. Not so, can feed on even less acceptable food
5. Races more common	5. Less common
6. Many modes of creation and maintenance of new races.	6. Less
7. Less complex digestion	7. Digestive fluids more complex
8. More asexual type of reproduction	8. Only sexual type except some parthenogenetic cases
9. Epiphytotics are easy to be created	9. Difficult
10. Breeding is comparatively easier	10. Resistance breeding is difficult

(ii) **Antibiosis:**   Painter (1936) proposed the term antibiosis for all those adverse effects on the life and life history of insects when resistant variety is used as food. Antibiosis may result in death of the first instar larvae and young nymphs, abnormal life cycle, abnormal physiology, lower fecundity and small size of insect. Resistance to brown plant hopper in rice varieties is due to antibiosis. New tools in biotechnology are adding another dimension by bringing in antibiosis gene from unrelated taxa like Bt gene from a fungus to develop cotton bollworm resistant Bt cotton.

(iii) **Tolerance:**   Compared to the first two mechanisms, tolerance is solely a host character. For insect resistance, tolerance has been related with good vigour and heterosis of host and replacement, regrowth and tissue repair by the host. For example, heterotic F_1 of two susceptible parents of sorghum is tolerant to chinch bug. Rice varieties are tolerant to stem borer by forming additional tillers to compensate yield losses. Inheritance of tolerance is complex in many cases and is supposed to be governed by polygenes.

9.2 Screening Procedure

Screening procedure for resistance depends on the nature of crop, life cycle of insect, its most damaging stage and the type of resistance being looked for. For example, in all defoliating insects egg masses or larvae are placed in leaf whorls. Egg masses are also released in stem borers like stem borer of maize. Egg laying females are released in caged plants as for many jassids and aphids. Besides these, insect outbreaks in some years take place, and survey of that area provides a chance to select resistant types. The screening material is also placed in heavily infested areas. The field population may be augmented by releasing artificially reared insects. For further details of screening see Panda (1979).

9.3 Breeding Procedure

The steps in breeding for insect resistance are similar to disease resistance, that is, screening and isolation of donors, incorporation of resistance and then development of the resistant variety. All those breeding methods as described with disease resistance are followed here also depending on the crop, donor and the insect-pests. For illustration, the example of rice in developing brown plant hopper resistance is given below.

Brown plant hopper (*Nilaparvata lugens*) sucks the cell sap and results in typical hopper-burn. Resistant donors like C022, Ptb 18, Rathuheeneti and Babawee which have respectively Bph 1, bph 2, Bph 3 and bph 4 genes for resistance are used in hybridisation. F_2 population is grown in small trays. Brown plant hoppers (BPH) are multiplied in cages and the third instar nymphs are released on the seedlings. Only resistant seedlings survive which are transplanted in the field. The same procedure is repeated in F_3 and F_4 generations. Thus a number of resistant progenies are advanced, tested for agronomic characters and yield performance to develop a high yielding and BPH resistant rice.

9.4 Breeding for Multiple Resistances

When resistance against more than one disease and pests are combined in one variety, it is known as multiple resistances. For example, rice varieties listed in Table 12.8 have multiple resistances. The breeding procedure is almost the same except in the use of multiple donors and screening procedure. For example, while breeding for multiple resistances against grassy stunt virus, bacterial blight, brown plant hopper and green leaf hopper, the following procedure may be used. About 400 F_2 seeds of multiple crosses involving four donors are germinated and inoculated with grassy stunt virus. Half of the seedlings would be susceptible and would be eliminated and the remaining 200 would be transplanted in the field and inoculated with bacterial blight. Half would again get eliminated and seeds from the remaining 100 would be harvested, grown and screened against brown plant hopper and green leaf hopper. By this screening 75 would be eliminated and F_2 progenies of only 25 plants would be advanced. Further selection for combined resistance and agronomically superior types in F_3, F_4 and F_5 generations would be done.

Table 12.8   Multiple resistance in some rice varieties compared to IR8 (S = Susceptible, R = Resistant, MR = Moderately Resistant) Khush (1977)

Variety	Disease and Insects					
	Blast	Bacterial blight	Tungro	Green leaf hopper	Brown plant hopper	Stem borer
IR 8	S	S	S	R	S	S
IR 20	SM	R	MR	R	S	MR
IR 26	MR	R	MR	R	R	MR
IR 30	MS	R	MR	R	R	MR
IR 36	MR	R	R	R	R	MS

10.   EXPLOITATION OF RESISTANCE GENES

Once incorporated, a number of schemes have been proposed to manage resistant genes to control diseases and pests.

10.1   Replacement of Succumbed Variety

The commonest scheme has been to replace a variety, which has gone susceptible with a resistant variety. This approach was adapted to control stem rust in Australia between 1938 and 1950 (Watson and Luing 1963). The same is being used now to control brown plant hopper in rice in Philippines. IR 26 rice variety with Bph 1 gene for resistance was released in 1973 but succumb to it in 1975. Then IR 36 and IR 38 carrying Bph1+bph 2 genes were released. By the time these succumb, varieties with Bph3 and bph 4 genes would be ready for release.

10.2   Advance Warning of Vulnerable Genes

A system has been developed by Australian and Indian wheat breeders and pathologist to issue advance warning to breeders not to use a resistance gene, which has gone susceptible in the advance testing. The procedure involves the continuous testing of resistant genes against new and more virulent races of pathogen.

10.3 Pyramiding

Watson and Singh (1962) suggested that 2, 3 or even more oligogenes be incorporated into a single genotype. Varieties with increasing number of resistant genes should be released in succession (1. 12.4). Thus with the available number of genes, the resistance can be used for a long time. Canadian oat breeders have adapted the same procedure against crown rust (Knott, 1974).

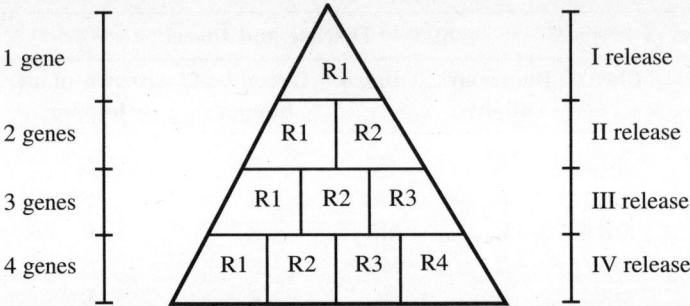

Figure 12.4   Pyramiding of 4 resistance genes R1, R2, R3 and R4

10.4 Recycling of Resistance Genes

Person (1967) suggested that there are limited resistant genes available. Therefore they can be rotated at some time interval. This would slow down the racial build up and would also avoid the creation of super races attacking all genes as feared in pyramiding.

10.5 Multilines

Borlaug (1953) clearly outlined the procedure to develop multiline cultivars to control stem rust of wheat. The resistance of multilines is due to a variety of mechanisms. A proportion of the initial inoculum falls on resistant lines, thus reducing the initial inoculum. Of the spores produced by infected susceptible lines, only a proportion will fall on other susceptible plants, thus reducing the rate of multiplication. In addition, the resistant plants act as barrier to dispersal. Further, Johnson and Allen (1975) suggested that resistance induced by infection with an avirulent race of a pathogen is effective against subsequent infection with a virulent race. Resistances of multilines function as synthesised horizontal resistance and does not break for a long time. Methods to develop multilines have been described in Chapter 6.

10.6 Multilineal Hybrids

Borlaug (1965) proposed that in place of fixed component lines, if hybrids are utilised, it will form a multilineal hybrid and will utilise heterosis also (see Figure 6.6) along with the advantages of multilines.

10.7 Gene Mosaics

If the epidemiological zones for a disease could be worked out, varieties with different resistant genes could be recommended in different geographical areas. This would enable keeping the level of primary inoculum lower. For example, Joshi and Nagarajan (1977) divided the Indo-Gangetic plains into four zones (Figure 12.5) and advocated the use of Lr 9, Lr 10, Lr 15, and Lr 19 genes to control brown rust of wheat. They suggested that each zone should have a specific gene or gene combination but not overlapping each other. This is a planned way to avoid monoculture, that is, cultivation of one variety over large geographical areas. Nelson (1973) called this type of geographical gene deployment as a geographical multiline. Contrary to monoculture where one virulent race may destroy the entire crop, in gene mosaics one virulent race may be limited to one area only and losses would be minimum.

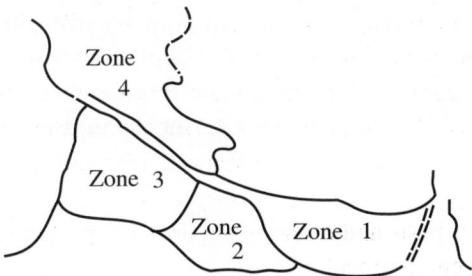

Figure 12.5 Four distinct brown rust epidemiological zones of north India.

10.8 Durable Resistance

Duration for which a variety remains resistant depends on the crop, the pathogen, nature of resistance and the location. In wheat rusts, the range in longevity is 1-10 years with an average of 5 years. In Texas state of USA, a wheat variety, Fox, remained resistant for one year only. Contrarily, the Russian variety, Lutescens 62, remained resistant for nearly 45 years. Apparently then, Lutescens 62 has good general resistance (Kilpatrick and Moseman, 1979). Durability of resistance is not a simple matter of number of resistant genes involved but also involves crop cultivation, rotation, length of vegetation, type of pathogen, alternative hosts and several other factors. However, keeping all other factors constant, generalised or horizontal or non-specific or partial or polygenic or uniform resistance are more durable (Pope and Dewey, 1979) than their counterparts. Even combining vertical and horizontal resistances in variety would be still better (Abdalla and Hermsen, 1971). Thus breeding for durable resistance has to proceed on lines one would breed for any quantitative character, say, yield. Care has to be taken that epiphytotics are created with a mixture of races and not by a pure one.

11.   ADVANTAGES OF RESISTANCE BREEDING

Control of diseases and insect-pests by resistance breeding has many plus points over other methods.

   (i) Once incorporated, resistant varieties go a long way in protecting crop losses due to pests and diseases.

   (ii) It is the cheapest way for control and does not add to the cost of production.

   (iii) Since resistance is built-in, it is ever ready unlike chemicals, which need to be applied in hours of need.

   (iv) It is the only control measure against many diseases where chemical or other control measures are not available like wilts, root-rots, rusts, soil borne, smuts, some nematodes and bacterial blight diseases.

   (v) From the cultivator's point of view, it is the surest and the best control measures and he has not to bother for any measure except the variety.

   (vi) It is safe for human life, unlike pesticides and fungicides, which are dangerous.

   (vii) It controls pests and diseases but does not poison or pollute the soil, water or environment.

   (viii) Success of control does not depend on external variation or preciseness of use, as in the case of other factors.

12.   LIMITATIONS OF RESISTANCE BREEDING

   (i) It becomes difficult to maintain balance between productive breeding and resistance breeding. There are varieties, which yield the highest but are susceptible.

   (ii) Linkage of resistant genes with genes of inferior qualities make it problematic to transfer.

   (iii) Self sterility in the host plant, which inhibits the transfer, is yet another problem posed in utilisation of some donors.

(iv) Suitable donors are not available against many pathogens such as ergot of bajra or sheath-rot of rice.

13. USEFUL REFERENCES

Allard, R.W. 1960. *Principles of Plant Breeding*. New York: John Wiley & Sons Inc.

Anon. 1974. Induced Mutations for Disease Resistance in Crop Plants. Vienna: IAEA.

Abdalla, M.M.F. and Hermsen, J.G.T. 1971. The concept of breeding for uniform and differential resistance and their integration. Euphytica 20: 351-361.

Borlaug, NE.E 1959. Use of multilineal or composite varieties to control airborne epidemic diseases of self pollinated crop plants. Proc. I int. Wheat Genet. Symp., Winnipeg, Canada.

Chopra, V. L. (Ed) 2001. *Breeding Field Crops*. Oxford & IBH Publ. Co. New Delhi, Calcutta.

Day, P. R. 1974. *Genetics of Host-Parasite Interaction*. San Francisco: W.H. Freeman & Co.

Flor, H.H. 1956. The complementary genic systems in flax and flax rust. Adv. Genet. 8: 29-54.

Johnson, R. and Allen D. J. 1975. Induced resistance to ruse diseases and its possible role in the resistance of multiline varieties. Ann. Appl.. Biol. 80: 1-5.

Joshi, L. M. and Nagarajan, S. 1977. Regional deployment of Lr genes for brown rust management. In: Genetics and Wheat Improvement. New Delhi: Oxford & IBH

Khush, G. S. 1977. Disease and insect resistance in rice. Adv. Agron. 29: 265-341.

Kilpatrick, R.A. and Moseman, J.G. 1979. 50 years of national and international wheat nurseries. Proc. V. Int. Wheat Genet Symp., 1978, New Delhi.

Lupton, F.G.H. and Johnson, R. 1970. Breeding for mature plant resistance to yellow rust in wheat. Ann Appl. Biol. 54: 697-700.

Nelson, R.R. 1973. *Breeding plants for Disease Resistance*. Penn. State Univer. Press.

Painter, R.H. 1952. *Insect Resistance in Crop Plants*. New York: the Macmillan Co.

Panda, N. 1979. *Principles of Host-plant Resistance to Insect-pests.* New Delhi: Hindustan Publ. Corpn.

Pope, W. K. and Dewey, W.G. 1979. Breeding strategies for handling quantitative resistance to plant diseases. Proc. V Int. Wheat Gene. Symp., New Delhi, 1978.

Raper, J. R. 1954. Sex in micro-organism Wash. DC: Amer. Soc. Adv. Sci.

Reddy, M. S .S. and Rao, M. V. 1977. Genetic classification and control of leaf rust pathogen of wheat in India. In: Genetics and Wheat Improvement. New Delhi: Oxford & IBH.

Schafer, J.F. 1971. Tolerance to plant diseases. Ann. Rev. Phytopathol. 9: 235-252.

Singh, S. S.; Sharma, R. K.; Singh, G.; Tyagi, B. S. And Saharan M. S. 2011. *100 Years of Wheat Research in India.* Directorate of Wheat Research, Karnal, Haryana, India, 281 pp.

Swaminathan, M. S. 1979. Recent trends in crop improvement in India. Proc. V Int. Wheat Genet. Symp. New Delhi, 1978.

Stakman, 1914. Study in cereal rusts: physiological races. Mime. Agric. Exp. Sta. Bull. 138.

Van der Plank, J.E. 1963. *Plant Diseases: Epidemics and Control.* New York: Academic Press.

---1968. *Disease Resistance in Plants.* New York: Academic press.

---1975. *Principles of Plant Infection.* New York: Academic Press.

Innovative Methods in Plant Breeding

Conventional (traditional) methods of plant breeding are still used mostly to develop new varieties of crop pants. However, in additions to those methods of generating genetic variability and advancing the population, a number of non-conventional methods have come up in the recent past. This has been possible because of development of innovative methods of generating and handling breeding material described in this chapter. These are lumped together as innovative methods of crop breeding.

1. WIDE HYBRIDISATION

In crop improvement programmes, the parents used in hybridisation are mostly different varieties of the same species i.e. intervarietal cross. In such cases F_1 hybrids are fully fertile, hybridisation and handling of segregating generations does not present any problem. But when desired characters are not available within the same species, crosses are made with parents from different species or genera. This is called wide hybridisation. Wide hybridisation includes interspecific (between species of same genus) e.g. *Oryza sativa* × O. *glaberrima*; or intergeneric (among different genera of same or different family) e.g. wheat with rye (*Triticum aestivum* × *Secale cereale*).

1.1 History

The first authentic record of a distant hybridisation for crop improvement was the production of a hybrid between Carnation (*Dianthus caryophyllus*) and Sweetwilliam (*Dianthus barbatus*) by Thomas Fairchild in 1717. Subsequently, a number of interspecific hybrids were produced but had only academic interest, no agricultural value. The intergeneric hybrid, **Raphanobrassica** is the amphidiploid out of a cross between radish (*Raphanus sativus*) and cabbage (*Brassica oleracea*). *Raphanobrassica* was produced with a view to combine the root of radish with the leaves of cabbage, but it inherited roots of cabbage and leaves of radish. *Triticale*, a manmade cereal, was first produced by Rimpau in 1890 by crossing wheat and rye. Due to its nature as a hardy crop in cooler climates, resistance to many disease and pests, tolerance to drought, Triticale established itself as a new crop.

Distant hybrids may be classified into two broad groups. The first group includes those that exhibit at least some fertility so that they can be maintained by selfing, intercrossing among themselves or backcrossing to the parental species. The second group consists of those hybrids that are completely sterile and have to be maintained clonally or by doubling their chromosome number and convert into fertile amphidiploids.

1.2 Barriers to wide hybridisation:

Barriers to wide hybridisation are many and variable from cross to cross. In some cases wide hybridisation may have no barriers, like any intervarietal cross e.g. cultivated tomato (*Lycopersicon escultentum*) × wild cherry tomato *L. pimpinellifolium*. But in cases having barriers, it may be due to:

(i) **Failure of fertilization:** Pollen may not germinate on the stigma or pollen tube may not grow through the style or may burst into the style or male gametes do not penetrate the embryo sac to fertilze the egg cell. Thus fertilization may fail due to various incompatibility reactions.

(ii) **Failure of zygote formation:** Even if fertilization takes place, zygote may not be formed due to various genetic or physiological reasons.

(iii) **Failure of zygote (embryo) development:** In many cases fertilization takes place and zygote is formed but the embryo does not develop beyond certain stage i.e. embryo abortion. This may be due to (a) lethal genes, as in *Aegilops*, (b) genotypic disharmony between the genomes of the two parental species, as in *Gossypium*,(c) chromosome elimination, as in *Hordeum* and *Nicotiana*, (d) cytoplasmic incompatibility, and (e) endosperm abortion. In some cases embryo may be formed but may be weak, poorly developed or endosperm underdeveloped. Such seeds upon maturity may look shrivelled and shrunk. Thus before it aborts,"embryo is rescued" around a week after fertilization and grown on special media to produce fully grown embryo or even small plantlets directly from the callus.

(iv) **Failure of seedling development:**   Some embryos do develop to full seed stage but seedlings fail due weakness, hybrid chlorosis or necrosis. Interspecific and intergeneric F_1 hybrids of wheat show both chlorosis and necrosis, which is often lethal. Chlorosis refers to improper chlorophyll development resulting in variable degrees of chlorophyll deficiency. In some cases it is possible to rescue the seedlings.

(v) **Sterility in distant hybrids:**   Even if cross has been successful and healthy seedlings have been obtained, still the hybrids may prove to be sterile. In case of asexually propagated plants like sugarcane, potato etc, the hybrid can be multiplied vegetativley. But in case of seed propagated crops, sterility poses a major hurdle. The sugarcane x maize hybrid produced at the Sugarcane Breeding Institute, Coimbatore, is completely sterile but being maintained by clonal propagation. The sterility of hybrids may be caused by cytogenetic, genetic or cytoplasmic factors.

(a) **Cytogenetic basis of sterility:**   Reduced chromosome pairing is common in distant hybrids which are sterile. In the extreme cases all the chromosomes may be present as univalents. The distribution of chromosome in such cases is irregular, and it leads to the formation of unbalance gametes resulting in partial to complete sterility. Interspecific crosses also show various kinds of abnormality in cell division. Fertility in such hybrids is improved by doubling their chromosome number i.e. by producing amphidiploids.

(b) **Genetic Basis of sterility:**   Chromosome pairing in some interspecific hybrids may be normal but show sterility. Sterility in such hybrids may be governed by genetic factor. Hybrids between cultivated foxtail millet (*Setaria italica*) and its wild relative *Setaria viridis* have normal chromosome pairing but are sterile indicating genetic factors.

1.3 Use of distant hybrids in plant breeding

Parents of distant hybrids differ in many genes and thus in F_2 very wide ranges of segregants are expected. Many of the segregants may look very different from those of the two parental species. Thus some of useful segregants may be developed as variety for cultivation. Additionally, new characters not available in the cultivated species of crop plants may also be introduced from the other species. Crops like wheat, oat, sugarcane, cotton, potato etc are allopolyploids have evolved though distant hybridisation.

(i) **Disease Resistance:** Disease resistance is by far the most common characteristic transferred against rusts in wheat, late blight in potatoes, bacterial wilt, bacterial canker, curly top virus , and mosaic virus in tomato, and against many insects and diseases in sugarcane. The potato famine in Europe during 1840s caused by late blight in which about a million Irish died as a result if this famine. A wild species of potato, *Solanum demissum* provided genes for late blight resistance, which served as a miracle saver of the potato crop. In cultivated wheat, genes Lr9 and Lr35 against leaf rust, Sr39 against stem rust, Pm12 and Pm13 against powdery mildew were transferred from various species of *Aegilops* genus. When all the cultivated varieties of rice (*Oryza sativa*) were found susceptible to devastating grassy stunt virus in 1970s, only resistance source was found in wild species from India, *Oryza nivara*, which is incorporated in all modern day rice varieties.

(ii) **Wider Adaptation:** Wild species have served as useful sources of genes for earliness and wider adaptation. Cold tolerance has been transferred from wild relatives to wheat, onion, potato, tomato, grape, rye and peppermint. Cold tolerant varieties adapted to much wider geographic region of the world.

(iii) **Mode of reproduction:** Genes affecting the mode of reproduction have been transferred from wild relatives to the cultivated species. Self incompatibility genes from *Brassica campestris* have been transferred to the self- compatible *B. napus* for the production of hybrid seeds.

(iv) **Yield:** Contrary to the general belief, wild relatives of many crop species are excellent sources of the much needed 'yield genes'. There are evidences from interspecific hybridisation in chickpea (*Cicer arietinum*). F_2 from the cross *C. arietinum* × *C. reticulatum* showed very large transgressive segregation for yield and yield components.

F_6 progenies showing up to 40 % improvement in 100 seed weight and 30 - 40 % increase in seed yield per plant over the *C. arietinum*.

(v) **New Varieties:** The cotton variety Varalakshmi is an interspecific hybrid between *Gossypium hirsutum* and *G. barbadense*. All present varieties of sugarcane (2n = 100 - 125) are complex interspecific hybrids involving *Saccharum officinarum* (2n = 80) and *S. spontaneum* (2n= 40 - 218) and other species of genus *Saccharum*. Current varieties of sugarcane are resistant to diseases and pests and also more productive. Bajra-Napier grass hybrids yield twice more fodder. Most widely cultivated rice variety Mahsuri is a hybrid derivative between *indica* and *japonica* subspecies of *Oryza sativa*.

(vi) **New Crop Species:** Interspecific hybridisation followed by chromosome doubling of the sterile hybrids produces allopolyploids that may become a new crop species like *Triticale*, or serve as bridging species to facilitate further wide hybridisation as in tobacco. F_1 of a cross between *Nicotiana tabacum* (2n = 48) and *N. glutinosa* (2n = 24) is sterile. But by doubling the chromosome number of the F_1, a fertile species *N. glutinosa* (2n=72) was created, which serves as a bridging species to transfer genes and chromosomes from *N. glutinosa* to *N. tabacum*.

1.4 Limitations of Wide Crosses

(a) F_1 hybrids of distant crosses generally exhibit variable sterility. Very rarely the F1 is fully fertile.

(b) Monogenic dominant genes are easily transferred than recessive or those governing quantitative characters.

(c) Some interspecific hybrids fail to flower as in soybean (*Glycine genus*) but flower only when grafted onto *Glycine max*.

(d) Outcome of the distant hybridisation is very poor agronomically unless superior variety is used as one of the parents.

(e) In some cases, the F_1 seed of the interspecies hybrids exhibit dormancy, like 5 - 10 years' dormancy in interspecific hybrids in groundnut (*Arachis*).

1.5 Achievements

Distant hybridisation has been most commonly used for the transfer of characters not found in the cultivated species like resistance to abiotic (heat, drought and submergence tolerance) and biotic (disease and insect-pests) stresses to cultivated species from its wild relatives. Examples are given of rust resistance in wheat, drought tolerance and perenniality in rice. Distant hybridization has also been used to transfer sterile cytoplasmic (WA = Wild Abortive) from wild *Oryza spontanea* to cultivated *Oryza sativa*. This gave the first ever male sterile (A line) used for producing hybrid rice seed. Many novel characters have been transferred to sugarcane from its wild relatives. Distant hybridisation has created a new cereal Triticale (*Triticale hexaploide*). Some other examples are *Raphanobrassica, Triticum - Agropyrin, Triticum- Aegilops* and *Festuca-Lolium* hybrids, etc.

2. TISSUE CULTURE

Plant part, tissue and cell culture can be utilized for plant improvement depending on the availability of methods to generate whole plant from these. Major parts like suckers of banana to cell- suspension are used to regenerate plants directly or via callus. Aborting embryos in the interspecific crosses can be rescued after a week of fertilization and then cultured to generate whole plant. In fertile F_1, further selection can be done using conventional methods of plant breeding. Anthers or other parts of the flowers can be used to generate callus and finally the seedlings. If pollen from the F_1 hybrid is cultured it can give haploid seedlings, which upon colchicine treatment can give a perfectly homozygous plants. This can save several years of time needed for fixation by pedigree method. This method has been used in breeding varieties in rice and several other crops, described under haploid breeding.

2.1 Plant generation

Shoot and root may be generated from the callus and such plantlets could be planted first under protected condition followed by under field condition. The resulting plants may produce seed which can be handled either by doubling the chromosome number (if haploid or sterile) or letting it produce seed for handling through selection. Tissue culture is used as an aid in plant breeding programmes variously. Plant can be generated using cell, pollen, anther, embryo, tissue, meristem, and other plant parts by using proper protocol either to rescue or to multiply it. Haploids or monoploids can be multiplied and prepared for doubling their chromosomes by colchicine to create diploid or amphidiploid. Aborting embryos could be rescued in the distant crosses to develop plant and handle using appropriate breeding technique. Under embryo culture, they produce seedlings directly without undergoing the cycle of seed maturation and dormancy. This shortens the breeding cycle.

2.2   Somatic embryogenesis

In cell culture, somatic embryos originate from single cell of the callus. They look like embryo and also proceed along the same stages of development as a seed i.e. induction, development, maturation and germination. But these embryos give out plantlets (seedlings) directly and may be haploid, diploid or polyploid. Depending on the tissues used, plantlets (seedlings) obtained from the somatic embryos can also be handled as a seed to start selection cycle.

2.3   Somaclonal Variation

The cells, somatic embryos or plantlets obtained from the cell culture also show heritable variations for qualitative characters. This is called somaclonal variation, and provide ample opportunity for selection to operate over the genetic variation present. Somaclonal variation may originate from chromosome structural changes (in polyploids), gene mutation etc. Exact reason is yet to be established for the origin of these variations. In case of potato, sugarcane and apple high occurrence of somaclonal variation has been recorded and used for developing new varieties. A mustard variety Jai Kisan was produced as somaclonal variant having 17% higher yield than its parent variety Varuna.

3.   BIOTECHNOLOGICAL METHODS

Biotechnology has opened ocean of opportunities for plant breeders. It enables transfer of genes from unrelated genera or family or even kingdom i.e. plant to animals and vice versa to crop plants. It also empowers breeders to practice precise selection and thereby saving time and money spent in the field.

3.1   Somatic hybridisation

Production of hybrid plants through fusion two different protoplasts (cytoplasm + nucleus) of two different plant varieties or species or genera is called somatic hybridisation, and hybrids thus produced are called **Somatic Hybrid**. In this case nucleus and cytoplasm of both the plants are present as only the cell wall was removed for fusing two protoplasts. In case nucleus of only one parent but cytoplasm of both is present in a hybrid, this is called **Cybrid** or **Cytoplasmic hybrid**. Protoplasts (cells without cell wall and enclosed by plasma lemma only) can be isolated from most plants and cultured to produce callus. Protoplasts of two neighbouring cells can be fused using Polyethylene Glycol (PEG). Somatic hybridisation involves (i) isolation of protoplast (ii) fusion of the parental protoplasts (iii) selection among somatic hybrids, and (iv) culture of the selected hybrid to generate plants.

Successful somatic hybrids have been obtained in potato + tomato, tobacco + carrot, tobacco + petunia and many other plant species. The technique works most when cytoplasmic factors are involved or when cytoplasm of one species has to be added to nucleus of the other. Using this technique male sterile cytoplasm (CMS) of *Brassica napus* has been transferred to *Brassica juncea*.

3.2 Gene transfer

Biotechnology has started entirely new era of plant breeding with new tools and new rules. Transfer of genes across the plant and animal kingdom is possible now. Using these tools plants have been created which are termed as "Transgenic Plants". Biotechnology is a developing science and the methods are being refined continuously. For common understanding the sequence of events in using the biotechnological tools in plant breeding are (I) introduce desired gene into the targeted plant, (II) integrate the introduced gene into the genome of the plant, (III) notice expression of the introduced gene in the transformed tissue / plant, (IV) regenerate the transformed tissue into plant, (V) test / observe the transformed plant for the introduced gene. Thus the first step (I) is most important and now following methods are available to transfer genes from other plants or animals into cultivated plants.

3.2.1 *Tumour Inducing (Ti) and Root Inducing (Ri) Plasmid*

The Ti and Ri plasmids are found in Agrobacterium and induce crown gall (tumour) or hairy root respectively in plants. These are used as carrier (vector) to introduce the "targeted gene" or "transgene" in the host plant. Then several copies of the "targeted gene" are first cloned (multiplied) using biotechnological tools in a bacterium called *Escherichia coli*. This *E. coli* is then introduced into the T-DNA region of the plasmid. This especially constructed plasmid then co-cultured with either protoplasm of the host or its leaf disc. By selective multiplication, the transformed cells are allowed to multiply and generate callus and then differentiate into plantlets. These plantlets are hardened and finally tested for the presence of the transgene to see if the transformation has taken place and is stable.

3.2.2 *Plant Viruses*

Plant viruses, both DNA and RNA viruses, are being targeted as vectors (carrier) of transgene into the desired host plant. Cauliflower virus is a DNA virus and the transgene is attached to it and converted into a suitable vector. The targeted host plant is then infected with this vector to transfer the transgene. Virus spread throughout the plant body and cells transform themselves with this foreign DNA (transgene). In many cases this method has not given stable transformation. These have been tried in wheat and cruciferous plants.

3.2.3 *Electroporation*

Multiple copies of the transgene is prepared and placed next to the protoplast of the recipient protoplast. A high voltage current (1,000 to 1,500 V/cm) is passed for a short period (for about 10 microseconds) in a specially designed equipment. This treatment creates micropore in the protoplast of the recipient plant. Transgenes enter through these pores and transform the protoplast. This technique has been used successfully in rice, maize, wheat, and tobacco.

3.2.4 *Direct DNA uptake*

Transient expressions of foreign genes have been observed in some tissues following its contacts with the source. Based on this observation, methods were devised to let plant cell, protoplast, pollen, embryo and whole seed take foreign genes. Polyethelyne Glycol (PEG) is known to change the cell membrane properties and thereby promotes the uptake of DNA through protoplast. PEG also increases the genetic transformation to the tune of 10-3. This technique has been demonstrated successfully in rice, wheat and sugarcane.

3.2.5 *Particle gun*

This is a very physical method and uses a particle gun to shoot the DNA in the protoplast, cell, tissue, embryo, and even leaf disc. Imagine firing a gun and the bullet entering in the body of a prey to spread the poison. In case of particle gun, the foreign DNA is coated on the small particles (0.5 to 5.0 μm) of gold or tungsten and shot at the target. The particles along with foreign DNA penetrate the cell and deposit it in the nucleus. Transformation takes place and the foreign DNA becomes integrated as demonstrated in tobacco, soybean, maize and wheat.

3.3 Uses of Gene transfers

Gene transfer using genetic engineering is done to cultivated plants for specific purposes like for conferring resistance against disease and insect-pests, resistance to viruses, resistance to herbicides, production of medicines etc. Gene transferred from alien species to resist diseases and pests have been a major success. A soil bacterium called *Bacillus thuringiensis* has Bt gene (*cry* gene) that codes for over 100 types of crystal protein (Cry protein). Cry proteins are toxic to a number of Lepidopterous pests like maize borer, cotton bollworm, pink bollworm, tobacco budworm, chickpea and pigeon pea pod borer, brinjal fruit borer etc.

Intestine of the larvae get cut once they feed on the plant tissue containing *cry* gene. Larvae die within few days. This gene has been transferred using genetic engineering to quite a few crop plants. India has used Bt cotton with success though due to human and environmental safety concerns, varieties with Bt gene in maize, brinjal and lady's finger have not been released for general cultivation in India.

Glyphosate is a herbicide which can kill any plant foliage it is sprayed on (total killer) and thus has been very widely used. Plant gene resistant to glyphosate was identified and transferred using gene transfer technology to crops like soybean, wheat, maize, tomato and tobacco to produce transgenic varieties. Now glyphosate if sprayed on these transgenic varieties of the crops would kill the weeds and any other plant growing in the field but not these transgenic varieties.

By transferring the genes that encode the surface antigen proteins from the disease causing pathogen to plants, it has been possible to breed plants that produce vaccine. Today several plant varieties of potato, banana, tobacco, maize etc are available that produce vaccines, which earlier were produced using animals. For example, a child can get immunized against polio virus disease just but eating vaccine producing banana. This is called edible vaccine.

Many novel plant varieties are being produced using gene transfer which was not possible with the traditional methods of breeding. Thus crossing barriers to transfer genes set by nature have been broken and theoretically it is possible to transfer genes from any organism to plants. Through genetic transformation transgenic plants with novel characters are being produced.

3.4 Molecular Markers & Marker Aided Selection

Selection is one of the most critical task a plant breeder performs in the segregating populations. Performance of the selected plant, made based on the morphological characters or "appearance of the plant" or "eyeball judgement", depends on the heritability of the character. If the heritability is high, the performance of the selected plant in the next generation would be good. This is due to the reason that the phenotype (P) is the interaction product of genotype (G) and environment (E) meaning P = G x E. To avoid this uncertainty, molecular markers have been developed to select the plants based on those i.e. selecting based on genotype (G) alone. Both major and quantitative genes could be marked and identified in a segregating population. Basically there are 3 types of molecular markers.

(i) **Isozyme:** Isozymes are variant forms of an enzyme detectable through electrophoresis. Based on the isozyme patterns, markers have been developed but not very precise and not in use currently.

(ii) **Restriction Fragment Length Polymorphism (RFLP):** RFLP denotes that a single restriction enzyme produces fragments of different lengths from the same stretch of DNA. Based on this RFLP maps are prepared, similar to conventional genetic map, which are specific to a particular chromosome or chromosome arm. Efficient protocols for preparing such maps have been developed. Thus RFLP maps developed for several plants like rice, maize, wheat etc are in use in selection as well as in varietal identification.

(iii) **Random Amplified Polymorphic DNAs (RAPDs):** RAPDS are detected by repeated cycles of DNA denaturation - renaturation DNA replication using a PCR (polymerase chain reaction) equipment. Using PCR equipment, RAPD markers are developed which are specific to a strain or character and are more reliable and cheap to use in breeding. RAPDs serve as very efficient tool in selecting desired plants out of segregating population. It also serves as a reference point in identifying a variety in a better way than morphological characters. A number of other markers systems are also available to aid in plant breeding.

Molecular markers that flank a gene determining a trait of agronomic interest can be used to track down the gene in segregating generations. This allows indirect selection for the trait irrespective of the stage of the development and prevailing environment. It is estimated that three backcrosses with marker aided selection can help recover the recurrent parent genotype to the same extent as achieved with six backcrosses under conventional phenotypic selection.

4. PARTICIPATORY BREEDING

Participatory breeding is nothing new to conventional breeding except that the segregating material is planted at several locations at the same time. Selection is done by a team of people, which may include farmers. In the next cycle material selected at one location is sent to the next one and thus the selected material gets exposed to new agro-climate and people. It is believed that the varieties coming out of such a process will be more stable in performance and closer to what the farmer desire. However not many bright example has come forward of such an approach although even CGIAR centres are experimenting with it.

5. USEFUL REFERENCES

Bhardwaj, D.N. (Ed.) 2012 Breeding of Field Crops. Agrobios (India), Jodhpur.

Harlan , J .R. 1976 . Genetic resource in wild relatives of crops. Crops.Sci.16: 329-333.

Chopra, V. L. (Ed) 2001. *Breeding Field Crops*: Theory and Practice. Oxford & IBH Publ. Co. New Delhi, Calcutta.

Razdan, M. K. 2003. An Introduction to Plant Tissue Culture. Oxford & IBH Publishing Co., New Delhi.

Simmonds, N.W.1979. *Principles of Crop Improvement*. Longman, London and NewYork.

Singh, B. D. 2005. *Biotechnology*. Kalyani Publishers, New Delhi.

Singh, B. D. 2009. *Plant Breeding Principles and Methods*. Kalyani Publishers, Ludhiana, New Delhi.

Stalker, H.T. 1980. Utilisation of wild species for crop improvement. Adv.Agron.33:111-147.

Stalker, H.T. and Murphy. J.P. (eds.) Plant Breeding in the 1990s C.A.B. International, Wallingford, U.K.

Organisation and Achievemnets of Plant Breeding

Agriculture was a State subject in pre-independence era and, even in an agricultural country like India, it was very much neglected. Breeding of all crops in most states was under the charge of one crop breeder or economic botanist. To investigate into the causes of food scarcity, the Royal Commission on Agriculture was appointed. This commission observed in 1928: "The basis of all agricultural progress is experiment... Most of the advanced nations of the world have developed and achieved agricultural revolutions because of sustained and systematic research. India which is basically an agricultural country and her economy is heavily dependent on it, agricultural research no doubt plays a pivotal role." As a result the Imperial (now Indian) Council of Agricultural Research was established on May 23, 1929 to promote, guide and co-ordinate agricultural research.

Earlier, the Imperial (now Indian) Agricultural Research Institute was established in 1905 at PUSA in Bihar. (Pusa stands for Phipps of USA). Mr.

Henry Phipps was an American out of whose 20,000 pounds donation the IARI was started. Later, in 1934 earthquake the massive building was destroyed (Figure 14.1) and IARI was moved to New Delhi in 1936. ICAR was reorganised in 1965 as autonomous body and commodity committees were abolished and all central research institutes were placed under its control.

1. INSTITUTIONS

Now there are 99 research and training organizations (4 deemed universities, 47 research institutes, 17 national research centres, 6 bureaux, and 25 project directorates) run by ICAR. In the following 29 institutions breeding work is conducted on various crops:

ICAR Research Institutions on plants:

1. Central Agricultural Research Institute, Port Blair (Andaman)
2. Central Arid Zone Research Institute, Jodhpur (Rajasthan)
3. Central Institute for Cotton Research, Nagpur (M. R.)
4. Central Institute of Arid Horticulture, Bikaner (Rajasthan)
5. Central Institute of Cotton Research, Nagpur (Maharashtra)
6. Central Institute of Sub-Tropical Horticulture, Lucknow (U. P.)
7. Central Institute of Temperate Horticulture, Srinagar (J & K)
8. Central Plantation Crop Research Institute, Kasaragod (Kerala)
9. Central Plantation Crops Research Institute, Kasargod (Karnataka)
10. Central Potato Research Institute, Shimla (H. P.)
11. Central Research Institute of Jute and allied Fibres, Barrackpore (W. B.)
12. Central Rice Research Institute, Cuttack (Odisha)
13. Central Soil Salinity Research Institute, Karnal (Haryana)
14. Central Tobacco Research Institute, Rajahmundry (A. P.)
15. Central Tuber Crops Research Institute, Thiruvanthapuram (Kerala)
16. Central Tuber Crops Research Institute, Thiruvanthapuram (Kerala)
17. Directorate of Research (Rice, Oilseeds, Sorghum at Hyderabad; Maize, Floriculture at New Delhi; Wheat at Karnal; Groundnut at Junagarh, Rapeseed & Mustard at Bharatpur; Onion & Garlic at Pune; Cashew at Puttur, Karnataka; Oil Palm at Pedavegi, A. P.).
18. Indian Agricultural Research Institute, New Delhi.
19. Indian Grassland and Fodder Research Institute, Jhansi (U. P.)
20. Indian Institute of Horticultural Research, Bangalore (Karnataka)
21. Indian Institute of Pulses Research, Kanpur (U. P.)

Figure 14.1   Remnants of Phipps Laboratory of Imperial Agricultural Research Institute (IARI), Pusa, Bihar. On the right is the massive *naulakha* building of IARI Pusa.

22. Indian Institute of Spices Research, Calicut (Kerala)
23. Indian Institute of Sugarcane Research Lucknow (U. P.)
24. Indian Institute of Vegetable Research, Varanasi (U. P.)
25. Jute Agricultural Research Institute, Barackpore (W.B.)
26. National Bureau of Plant Genetic Resources, New Delhi
27. National Research Centre (Banana at Trichi, Litchi at Muzaffarpur, Citrus at Nagpur, Grape at Pune, Pomegranate at Solapur, Seed Spices at Ajmer (Rajasthan)
28. Sugarcane Breeding Institute, Coimbatore (T. N.)
29. Vivekanand Parvatiya Krishi Anusandhan Sansthan, Almora (Uttarakhand)

The turning point in Indian agriculture came when, based on the Land Grant college pattern of USA, the first agricultural university was established at Pantnagar in 1960. This was followed by establishing a chain of agricultural universities in most of the states, now numbering 67 and transferring all agricultural researches to them. These universities are listed below:

A. State / Central Agriculture Universities:

1. Acharya N G Ranga Agricultural University, Rajendranagar, Hyderabad (A.P.) 500030
2. Anand Agricultural University, Anand (Gujrat) 388110
3. Assam Agricultural University, Jorhat (Assam) 785003
4. Bidhan Chandra Krishi Vishwavidyalaya, PO Krishi Vishwa Vidyalaya, Mohanpur, Nadia, (W.B.) 741252.
5. Bihar Agriculture University, Sabour, Bhagalpur (Bihar) 813210
6. Birsa Agricultural University, Ranchi (Jharkhand) 834006
7. Central Agricultural University, Imphal (Manipur) 795004
8. Chandra Shekhar Azad University of Agriculture and Technology, Kanpur (U.P.) 208002
9. Chaudhary Charan Singh Haryana Agricultural University, Hissar (Haryana) 125004
10. Chaudhary Sarwan Kumar Himachal Pradesh Krishi Vishwavidyalaya, Palamapur, (H.P.) 176062
11. Dr. Balasaheb Sawant Konkan Krishi Vidyapeeth, Dapoli, Ratnagiri (Maharashtra) 415712
12. Chhattisgarh Kamdhenu Vishwavidyalay, Anjara, Durg (Chhattisgarh)
13. Forest Research Institute, New Forest, Dehra Dun (Uttarakhand)248195.
14. Dr Punjabrao Deshmukh Krishi Vidyapeeth, Krishi Nagar, Akola (Maharashtra) 4441004.

15. Dr Yashwant Singh Parmar University of Horticulture and Forestry, Nauni, Solan (H.P.) 173230
16. YSR Horticultural University, Venkataramannagudam, Tadepalligudam (A.P.) 534101
17. Govind Ballabh Pant University of Agriculture and Technology, Pantnagar (Uttarakhand) 263145
18. Guru Angad Dev University of Veterianry and Animal Sciences, Ludhiana (Punjab) 141004
19. Indira Gandhi Krishi Vishwavidyalaya, Krishak Nagar, Raipur (Chhattisgarh) 492006.
20. Jawaharlal Nehru Krishi Vishwavidyalaya, Jabalpur (M.P.) 482004.
21. Junagarh Agriculture University, Motibaug, Junagarh (Gujrat) 362001
22. Karnataka Veterinary Animal and Fisheries Sciences University, Nandinagar, Bidara (Karnataka) 585401
23. Kerala Agricultural University, Vellanikkara, Thrissur (Kerala) 680656
24. Kerala Veterinary & Animal Sciences University, Pattom, Thiruvanthapuram (Kerala) 695004
25. Lala Lajpat Rai University of Veterinary and Animal Sciences, Hissar, (Haryana)
26. Maharana Pratap Univeristy of Agriculture and Technology, Udaipur (Rajsthan) 313001
27. Maharashtra Animal Science and Fishery University, Nagpur, (Maharashtra)
28. Mahatma Phule Krishi Vidyapeeth, Rahuri, (Maharashtra) 413722
29. Manyawar Kanshiram Ji University of Agriculture and Technology, Banda (U.P.) 210001
30. Marathwada Agricultural University, Parbhani (Maharashtra) 431402
31. Nanaji Deshmukh Veterinary Science University, Jabalpur (M.P.) 482001
32. Narendra Dev University of Agriculture and Technology, Kumarganj, Faizabad (U.P.) 224001
33. Navsari Agricultural University, Vijalpore, Navsari (Gujrat) 396450
34. Orissa University of Agriculture and Technology, Bhubaneswar (Odisha) 141004
35. Punjab Agriculture University, Ludhiana (Punjab) 141004
36. Rajasthan University of Veterinary & Animal Sciences, Bikaner (Rajasthan) 334006
37. Rajendra Agricultural University, Pusa, Samastipur (Bihar) 848125
38. Rajmata Vijai Raje Scindia Agricultural University, Gwalior (M.P.) 474002

39. Sardar Vallabh Bhai Patel University of Agriculture & Technology, Modipuram, Meerut (U.P.) 250110
40. Sardar Krushinagar - Dantiwada Agricultural University, Sardarkrushinagar, Dantewada, Banskantha (Gujrat) 385506
41. Sher-e-Kashmir University of Agricultural Sciences and Technology, Srinagar (J. & K.) 190001
42. Sher-e-Kashmir University of Agricultural Sciences and Technology, Railway Road, Jammu (J. & K.) 190001
43. Sri Venkateswara Veterinary University, Tirupati, Chittoor (A.P.) 517502
44. Swami Keshwanand Rajasthan Agricultural University, Bikaner (Rajasthan) 334006
45. Tamil Nadu Agricultural University, Coimbatore (T.N.) 641003
46. Tamil Nadu Fisheries University, Nagapattinam, (T.N.) 611002
47. Tamil Nadu Veterinary and Animal Sciences University, Chennai (T.N.) 600051
48. University of Agricultural Sciences, Dharwad (Karnataka) 580005
49. University of Agricultural Sciences, Hebbal, Bangalore (Karnataka) 560065
50. University of Agricultural Sciences, Raichur (Karnataka) 584102
51. University of Horticultural Sciences, Navanagar, Bagalkot, (Karnataka) 587102
52. U.P. Pandit Deen Dayal Upadhyay Pashuchikitsa Vigyan Vishwidyalaya Evam Go Anusandhan Sansthan, Mathura, U.P.) 181001
53. Uttar Banga Krishi Vishwavidyalaya, Pundibari, Cooch Bihar (W. B.) 736165
54. University of Horticultural Sciences, Navanagar, Bagalkot (Karnataka) 587101
55. Uttarakhand University of Horticulture & Forestry, Pauri Garhwal (Uttarakhand)
56. West Bengal University of Animal and Fishery Sciences, Belgachia, Kolkata (W.B.) 700037

B. Deemed-to-be Universities in Agriculture

1. Central Institute of Fisheries Education, Seven Bungalow, Mumbai (Maharashtra) 400061
2. Indian Agricultural Research Institute, Pusa, New Delhi 110012
3. Indian Veterinary Research Institute, Izatnagar (Uttarakhand) 243122
4. National Dairy Research Institute, Karnal (Haryana) 132001

5. Sam Higgins Institute of Agriculture & Technology, Naini, Allahabad (U.P.) 211007

C. Central Universities with Agriculture Faculty

1. Aligarh Muslim University, Aligarh (U.P.)
2. Banaras Hindu University, Varanasi (U.P.)
3. Nagaland University, Medizipherma (Nagaland)
4. Vishwa Bharti, Shantiniketan (W.B.)

Most of these universities have more than one campus and a chain of research stations where breeding researches on various crops are done. Researches done by the students also form an integral part of breeding work at these universities.

2.   CO-ORDINATED RESEARCH PROGRAMME

The other turning point in breeding researches started in 1964, when the co-ordinated programmes on the pattern of maize were initiated in other crops. The All India Coordinated Research Project (AICRP) started with the philosophy of putting the agronomist, pathologist and entomologist with the plant breeder of the entire country on one platform. This was a significant positive departure from the past when breeding researches were done in isolation. Thus as a variety developed now by a plant breeder is tested in multi-locations by breeders, agronomists, pathologists, entomologists and others of the country before its release. Now there are co-ordinated projects in all major crops including fruits and vegetables, under the charge of a Project Director/Project Co-ordinator. For each crop, the country is divided into zones, which make the basis of varietal releases. Currently there are 59 AICRP operating in India.

The 1960's was also the era of major reorganisations in ICAR, and dwarf high yielding varieties and thus may be called an era of 'agricultural renaissance'. The total food grain production during 1949-50 was 5,4,21,000 tons, which rose to 10,84,22,000 tons during 1970-71. This breakthrough in production mainly due to high yielding varieties was called 'Green Revolution", and Dr. Norman Borlaug was named as "Father of Green Revolution and received Nobel Prize.

3.   ACHIEVEMENTS IN SOME CROPS

A brief history and achievements in some crops are described briefly below.

3.1   Rice (*Oryza sativa*)

Watt (1891) initiated botanical studies on rice in India. In an organised way, rice breeding was started in 1911 at Dacca (now in Bangladesh). Various states like Tamil Nadu (1913), Bihar, and Orissa (1914) U. P. and others started breeding work gradually.

It was in 1929 that the ICAR sponsored and aided special research schemes in states. The Central Rice Research Institute, Cuttack, was established in 1946. Prior to 1960, as many as 430 improved varieties were developed mostly by pure line selection. Only 27 were developed by hybridisation. The concept of ideal plant type was conceived and breeding started for dwarf high yielding type after the introduction of Taichung (Native) 1 in 1965 from Taiwan through Philippines. In 1965, the All India Co-ordinated Rice Improvement Project with its headquarters at Hyderabad started functioning. There are seven main agroclimatic zones each co-ordinated by a zonal co-ordinator. At present rice breeding is being done at 23 agricultural universities and other institutions numbering around 100.

Under the japonica x indica hybridisation programme, ADT 27 was bred in Tamil Nadu and Malinja and Mahsuri were introduced from Malaysia. Since 1965, a total of 20 high yielding varieties were released by the Central committee and 130 through various state committees. Some of the popular ones have been listed in Table 14.1. Out of these, 3 are mutants, 11 direct introductions, 1 back cross (Sabarmati), hybrids, 8 pure line selections and the rest are out of hybridisation handled through the pedigree method.

Rice, being the traditional crop, had trend setting breeders like Dr. R.H. Richharia, Dr. K Ramiah, Dr. N. Parthasarthy, Dr. S.V.S. Shastry, Dr. M. Mahadevappa, Dr. V. P. Singh, Dr. R. C. Chaudhary, Dr. E. Siddique, Dr. A. K. Singh and others.

3.2   Maize (*Zea mays*)

Maize was introduced in India some 200 years back, and much breeding work has been done here. The momentum came when the Inter Asian Corn Improvement Programme started.

Maize was the first crop in which co-ordinated research programme started in 1957 with headquarter at IARI, New Delhi. There are main zonal centres at Srinagar, New Delhi, Pantnagar and Hyderabad. The activities of the Inter-Asian Corn Improvement Programme and the Rockefeller Foundation in India brought momentum and the first hybrid maize Ganga 1 was released in 1961. Earlier there were some locally improved varieties like KT 41 OF U. P., Jaunpuri, Kalimpong, etc. Details of other varieties have been given in Table 13.2. Maize has the privilege of having breeders like Dr. N.L. Dhawan, Dr.

Table 14.1 Some popular high yielding varieties of rice released by Central Variety Release Committee (CVRC) and State Variety Release Committees (SVRC) and method of their breeding

Variety	Parentage	Year	Breeding method
C.V.R.C.:			
1. IR 8	DGWG/Peta	1966	Introduction from IRRI Philippines
2. Jaya	T (N) 1/T 141	1968	Pedigree
3. Jagannath	T 141 Mutant	1969	Mutation
4. Sabarmati	(T(N)1/Basmati 370)* Basmati 370	1970	Back cross
5. PRH10	Basmati A line/Pusa Sugandh 2	2001	Hybrid using CMS system
6. Shivam	CR314-5-10 mutant	2010	Pedigree
7. NDR2065	Pant Dhan 4/Saket 4/NDR2018	2006	Mutation
		2011	Pedigree
Andhra Pradesh:			
1. Tella Hamsa	HR 12 / T (N) 1	1971	Pedigree
2. Surekha	IR 8 / Siam 29	1976	Pedigree
3. Early Sambha	RNRM-7 from Mahasuri	2001	Mutation
4. Deepti	MTU4870	2001	Pedigree
Bihar:			
1. Sita	IR 8 / IR 12-178-2-3	1972	Selection from introduced segregating material
2. Sujata	BG90-2	1984	Introduction from Sri Lanka
3. Rajendra Mhasuri-1	BR51-46 / Mahsuri	1996	Pedigree
4. Birsa Vikas Dhan 110	T(N)1 / Brown Gora 23-19	2005	Pedigree
Gujarat:			
1. GAUR 1	Zinya 31 / IR 9-29	1973	Pedigree

2. GR-8	Selection local germplasm	2003	Pureline selection
Jammu & Kashmir:			
1. K 84	T 65 mutant	--	Mutation
2. K 78-13	Shinei / Ch. 971	--	Cross of two introductions
3. Chenab	SKAU-23	1996	Pedigree
Karnataka:			
1. Manila	C 4-63 selection	1970	Introduction from Philippines
2. Intan	Intan	1975	Introduction from Indonesia
3. Tunga	IET13901	2006	Pedigree
Madhya Pradesh:			
1. Garima	Cross 4 / T (N) 1	1975	Pedigree
2. Poornima	R-281-PP-31-1	1997	Pedigree
3. Shymla	R259-WR-37-2	1997	Pedigree
Maharashtra:			
1. Surya	BG 79 / IR 400-28-4-5	1973	Pedigree
2. Karjat 14-7	IR 8 / Zinya	1975	Pedigree
3. Phule Swarna	RHR-87015	2001	Pedigree
4. Sahyadri	KJTRH-12	2007	Hybrid
Orissa:			
1. Kalinga 2	Dunghansali / IR 8	1973	Cross with Hungarian introduction
2. Ajay	Hybrid	2006	Hybrid using CMS system
3. Jogesh	CR544-1-3-4 / NDR1008	2007	Pedigree
Punjab:			
1. Hybrid mutant 95	Jhona 349 / T (N) 1	1972	Hybrid mutation
2. Palman 579	--	1972	Introduction from Philippines

3. PR 106	IR 8 / Peta/5 // Belle Patna	1978	Multiple cross, introduced from Philippines
4. Punjab Basmati	Basmati 386 / Super Basmati	2008	Pedigree
Rajasthan:			
1. Chambal	IR 8 / NP 130	1975	Pedigree
2. Mahisugandha	BK70 / Basmati 370	1995	Pedigree
3. Vagad Dhan	M63-83 / Cauvery	1999	Pedigree
Tamil Nadu:			
1. Vaigai (CO 37)	T (N) 1 / CO 29	1974	Pedigree
2. TKM 9	TKM 7 / IR 8	1978	Pedigree
3. ADT 49	CR1009 / Jeeraga Samba	2012	Pedigree
4. ADT50	BPT5204 / CR1009	2012	Pedigree
Uttar Pradesh:			
1. IR 24	IR 8 / IR 127-2-2	1972	Introduction from IRRI
2. Prasad	IR747B2-6-3 / IR579-48	1978	Selection from introduced segregating germplasm
3. Jalmagan	Selection from Barho	1978	Selection from local variety
4. Sarjoo 52	T(N) 1 / Kashi	1982	Pedigree
5. Malviya Dhan 36	Mahsuri mutant	1997	Mutation
6. Malviya Sugandh 4-3	Mutant of Lanjhi	2009	Mutation
7. Kalanamak KN3	Kalanamak selection	2010	Pureline selection from Kalanamak
8. NDR2065	Pant Dhan 4/Saket4//NDR2018	2011	Pedigree

Boshi Sen, Dr. R.L. Paliwal, Dr. Joginder Singh, Dr. V.L. Asnani, Dr. V. P. Mani, Dr. N. N. Singh, Dr. M. P. Gupta and others.

In addition to the above varieties, 15 composite varieties and two hybrids have been recommended by various State Departments of Agriculture. Some of the important ones are for U.P. and Rajasthan (Moti), Bihar (Rajendra makka 1, Diara composite, Laxmi) Andhra Pradesh (Amber Pop), and Madhya Pradesh (Chandan 1, Chandan safed 2, Chandan 3).

The development and release of high yielding composites have provided considerable relief in overcoming the problem of hybrid seed purchase every year by farmers. The composite varieties give about 90 per cent of yield of the best hybrids. But technological changes made single cross hybrids affordable and thus single cross hybrids getting popular.

Table 14.2   Some maize varieties (hybrids and composites) released since 1961

Variety	Parentage	Year	Remarks
Hybrids:			
1. Ganga 1	(CM 101 × CM 102) × (CM 100 × CM 200)	1961	Double cross, flint
2. Ranjit	(CM 103 × CM 104) × (CM 202 × CM 106)	1961	Double cross, bold flint
3. Hi-Starch	(CM 400 × CM 300) × CM 601	1963	Double top cross, white dent.
4. HQPM 7	HHKI-193-1 × HKI-161	2008	Single cross
5. HM 12	KI-1344 × HKI-1378	2012	Single cross
6. CO 6	UMI 1200 × UMI 1230	2012	Single cross
Composites:			
1. Jawahar	A1 × Ant Gr. 1	1967	Yellow, semi flint
2. Kisan	J 1 × Coastal Tropical flint	1967	Yellow, semi flint
3. Hemant	Composite of Dholi 7744	1986	Yellow, flint, early
4. Megha	Composite of local Punjab germplasm	1993	Yellow, semident, early
5. Vivek Sankul Makka 7	Synthetic of 7 varieties	2009	Orange, flint, early

3.3   Wheat (*Triticum aestivum*)

Wheat breeding was started in 1906 by Mr. Albert Howard and his wife Mrs. Gabrille at IARI, (Bihar) in a systematic way. Wheat breeding in India, which has passed three distinct phases from 1906, did not boost up the yield significantly. The first phase yielded pure line varieties like NP4, PB8, K 13, Bansi, etc.

The second phase, which aimed at bringing different characters together through hybridisation, gave rise to varieties like NP 52, C 518, C 591, Niphad, etc. The third phase, primarily aimed at producing disease resistant varieties, yielded NP 700, and NP 800 series of Dr. B. P. Pal.

The era of semi-dwarf wheats started with the discovery of the Norin 10 dwarfing gene. Dr. Norman E. Borlaug and his associates incorporated this gene into spring wheats and some of this material was brought to India in 1961-62. Tests conducted on them during 1963-66 proved their superiority and as a result Sonora 64 and Lerma Rojo 64 were introduced from Mexico in large quantities. However, their grain was red and in 1967, the white grained Kalyan Sona and Sonalika were released.

Drs. Boshi Sen, A. B. Joshi, S.M. Sikka, M. S. Swaminathan, K. S. Gill, M. S. Kohli; M. V. Rao, H. N. Pandey and the young generation of wheat breeders set the record of elevating the genetic potential of wheat varieties. All this could happen due to the dynamic heat breeders' team in the country. The co-ordinated programme started in wheat in 1964 with its headquarter at IARI, New Delhi. Now there are nine zones in the country and 41 institutions participating in it. Some of the popular varieties are listed in Table 14.3. The first multiline variety to be released was Bithur in U. P.

Table 14.3   Some recently released and popular varieties of wheat

	Name	Parentage	Year	Remark
1.	Sonora 64	Yaktana 54 × Yaqui	1965	Introduction
2.	Kalyan Sona	Penjamo 62 sib × Gabo 55	1967	Selection from S 227
3.	Sharbati Sonora	Sonora 64 mutant	1968	Mutation
4.	Sonalika	II 54-388-An (Yy.54) × NIOB × Lerma Rojo = S308	1967	Reselection from S 308
5.	UP 262	S 308 / BJ68	1973	—
6.	Bithur	–	1979	First multiline in India
7.	PBW 343	ND/VG9144//KAL/BB/ 3/YCOS/4/Vee3S "S"	1995	Wide adaptability
8.	HUW 510	HD2278/HUW234//DL230-16	2001	Plain Zone
9.	DBW 39	Attila / Hui	2010	Pedigree
10.	HD 2967	Ald/Coc/Uresh/HD2160M/ HD2278	2011	Pedigree
11.	COW – 2	Mutant from NP 200	2012	Mutation
12.	HD3043	PJN/BOW/Opata*2/2/ CROC-1/ACS224/Opata	2012	Pedigree

3.4 Pearl Millet (*Pennisitum glaucum*)

Pearl millet or Bajra is an introduced crop from Africa. Hence not much breeding work was done in the country. Some work was taken up by the states in the early part of this century by way of introduction and selection, out of introductions from Africa, Jamnagar Giant and Improved Ghana were selected. Tift 23 A was utilised as a male sterile parent for the production of hybrid bajra. Since Tift 23A has gone susceptible to downy mildew and ergot, new male sterile lines have been developed using mutation breeding and back cross. These are M S 5071A, MS 521, Pb 111A, M S 5054 and MS 5141. Some hybrids are listed in Table 14.4. Hybrids from serial Nos. 1 to 5 are out of cultivation due to their susceptibility to downy mildew.

Table 14.4 Bajra hybrids and composites, their parentage and year of notification

	Name	Parent	Years of release	Remark
1.	HB-1	Tift 23 A × Bil 38	1965	Hybrid
2.	HB-2	Tift 23 A × J. 88	1967	Hybrid
3.	HB-3	Tift 23 A × J. 104	1968	Hybrid
4.	NBH-5	MS 5071 A × K 559-85	1975	Hybrid
5.	PHB-14	Pb 111 A × PIB 228	1975	Hybrid
6.	Parbhani Sampada	Composite of 8 inbreds	2005	Composite
7.	Saburi	RRRB5A / RHRBI 458	2005	Hybrid
8.	Mandor Bajra	Composite of 12 elite lines	2011	Composite
9.	ABPC4-3	Composite of 8 inbreds	2012	Composite
10.	PKV Raj	BMS5-23A / BR333	2012	Hybrid

3.5 Sorghum (*Sorghum bicolour*)

Being an introduced crop from Africa, the initial breeding work was concentrated on introduction and selection. Mass selection, pure line and recurrent selections were also applied. An accelerated Hybrid Sorghum Breeding Project was initiated in 1960 to breed hybrid varieties. The first hybrids were released in 1964-1965 in the name of CSH-1, CSH-2. Co-ordinated Sorghum Improvement Project started in 1969, with its headquarter at Hyderabad. The first hybrid CSH-1 was developed by Dr. S. S. Athwal. Under the able guidance of Dr. N. Ganga Prasad Rao, Sorghum breeding progressed the recent release includes CSV-27 (Figure 14.2).

Figure 14.2 Recent hybrid sorghum CSV-27 (Courtesy: Dr. J. V. Patil)

Table 14.5 Some varieties and hybrids of sorghum (jowar)

	Variety/Hybrid	Parentage	Year	Remarks
1.	CSH 1	CK 60 X IS 84	1964	Early, first hybrid
2.	Swarna	IS 324	1968	Mass selection
3.	CSH-4	1036 A X Swarna	1970	Early hybrid
4.	Parbhani Sweta	Selection from GDLP 34-5-53	2005	Mass selection
5.	Bundela	SPV946 X KSE33	2005	Pedigree
6.	Pusa Chari 9	Sel. IS4870	2008	Mass selection
7.	AKSSV 22	S-171 X HES-13	2009	Pedigree
8.	CSV 27	GJ 38 X Indore 12	2010	Single cross hybrid
9.	Kinneria	Moti X D71258	2010	Pedigree
10.	Pratap Chari 1080	UCSR X SPV	2011	Pedigree
11.	CO 30	APK 1 X TNS291	2012	Pedigree
12.	Phule Panchami	Selection from local Landrace	2012	Mass selection

3.6 Sugarcane (*Saccharum spp.*)

Three species of *Saccharum* genus are cultivated; *S. officinarum* is thick stem chewing type; *S. barberi* and *S. sinensis*, thin stem commercial cane type. Modern sugarcane varieties are complex interspecific hybrids. Sugarcane breeding started in India in the beginning of the 20th century under the able guidance of Dr. C.A. Barber. There are two main institutes where breeding work is conducted, the sugarcane Breeding Institute, Coimbatore, and the Indian Institute of Sugar Cane Research, Lucknow.

The Sugarcane Breeding Institute was started in 1912. In addition, the All India Co-ordinated Research Project on Sugarcane Improvement has many centres in the country. Since uniform flowering and viable seed setting takes place in Coimbatore, the crossing work in done mainly there (see Figure 8.4). Seedlings are grown at a number of locations in the country including Pusa, Lucknow, Shahjahanpur, Pantnagar and Karnal in north India. All crosses made at Coimbatore bear a prefix CO, and cross number (Table 13.6). The place of subsequent selection is added later; for example, COS 410 means cross number 410 made at Coimbatore and later selections made at Shahjahanpur. The BO prefix is used for crosses made at Pusa in Bihar. Some varieties were released bearing BO numbers.

Table 14.6   Some popular sugar cane varieties, their parentage and year of notification

	Variety	Parentage	Year	Remark
1.	CO 1148	P 4383 X CO 301	–	High yield, resistant red rot & salinity good ratooning ability
2.	BO 99	CO 1207 X BO 43	-	Resistant to water logging
3.	Sarayu	CO312 / CO6806	2000	Hybridisation, clonal selection
4.	Haryana 92	CO7704 G.C.	2001	Clonal selection
5.	Gandak	BO91 G.C.	2001	Clonal selection
6.	CO Pant	Outcross Co Pant 84212	2007	Outcross clonal selection
7.	Sweety	COS8119 / CO62198	2007	Hybridisation, clonal selection
8.	CO 238	COLK / CO775	2009	Hybridisation, clonal selection
9.	SNK 004	CO740 / COA7602	2010	Hybridisation, clonal selection
10.	CO92005	COC671 / COT8201	2011	Hybridisation, clonal selection
11.	Haryana ganna	COH70 / COS510	2012	Hybridisation, clonal selection
12.	Karan 9	COS8436 / CO89003	2012	Hybridisation, clonal selection

3.7   Potato (*Solanum tuberosum*)

Unconscious potato breeding in India started with its introduction in 1615 from Europe. Acclimatisation of introduced varieties continued for over 300 years by various states and other research organisations. A few hundred varietal names were under cultivation but all of them belonged to 16 adapted ones. A potato breeding station was started in 1935 at Shimla, which became the Central Potato Research Institute later. Now there are eight regional and number of sub centres of potato breeding in the country. Potato is an asexually (tuber) propagated crop but flowers under the long day of summer in hilly temperature and sets fruits also. Therefore crossing is done only at Shimla and Darjeeling. The crossed seeds are sent to Jalandhar, Patna and a few more centres where seedlings are raised and selection is made for the local conditions. For example, at Darjeeling, selection is done for wart resistance. The first batch of varieties was released after 1966 out of crosses made in India with a prefix Kufri. Some of the popular ones are described in Table 14.7. Now with the 'seed plot technique' the new varieties are multiplied quickly in the plains also and their spread is hastened.

Due to the high yielding varieties, the yield of potato has been quadrupled from 6.6 to 23.68 tons per hectare respectively for 1949 and 2012 (FAOST AT 2013).

3.8 Cotton (*Gossypium spp*)

Four species of cotton *Gossypium hirsutum*, *G. herbaceum*, *G. arboreum*, and *G. barbadense*, and their hybrids are under cultivation in India. Cotton improvement work started in India in 1790. In the following 100 years various types from America and Egypt were introduced but they were not successful due to their susceptibility to diseases and pests. In 1906 Cambodia cotton was introduced, and being resistant to jassids, this became very popular in south India.

In 1917 the Indian Cotton Committee was established to breed long staple cotton. In 1921 Indian Central Cotton Committee was established which provided assistance and guidance to state research stations. It was abolished in 1965 and the cotton research put under ICAR. Now Central Cotton Research institute has been established in Nagpur to co-ordinate and give a lead in cotton research.

The All Indian Co-ordinated Cotton Improvement Project started in 1967 and to date there are 30 centres in various agroclimates. A number of varieties have been released and are in cultivation (Table 14.8). A real breakthrough in cotton production was achieved with the development of hybrid cotton 'HP' at Surat. This was the first hybrid cotton in the world to go into commercial production. This was followed by an interspecific hybrid Varalakshmi developed at Dharwar. The development and release of high spinning cottons Sujata and Suvin is a landmark in cotton improvement. Suvin is capable of spinning 120 counts which is as good as the best Egyptian cotton.

Table 14.7 Some popular varieties of potato, their parentage and year of notification

	Variety	Parentage	Year	Remark
1.	Kufri Alankar	O.N. 2090 X Kennebec	1967	Photo-insensitive
2.	Kufri Sindhuri	Kufri Red X Kufri Kundan	1967	Waxy, frost tolerant
3.	Kufri Chandra Mukhi	Kufri Kuber X 54485	1968	Male sterile
4.	Kufri Deva	Craigs Defiance X Phulwa	1973	Storage even at room temp.
5.	Kufri Bahar	E 3797	1979	-
6.	Kufri Badshah	Kufri Jyoti X Kufri Alankar	1982	Hybridisation, clonal selection
7.	Kufri Sherpa	Ultimus X Adina	1984	Hybridisation, clonal selection
8.	Kufri Kanchan	SLB/Z405 (A) X Pimpernel	2001	Hybridisation, clonal selection
9.	Kufri Chipsona-3	–	2006	Special purpose
10.	Kufri Sadabahar	MS/81-145 X PHF/-1545	2008	Hybridisation, clonal selection
11.	Kufri Frysona	MP/98-71	2010	Clonal selection

Table 14.8 Some cotton varieties notified for cultivation and breeding methods used

	Variety	Parentage	Year	Remark
1.	Amaravati	E/L. 389 / 11969	1980	Pedigree
2.	MCU 7	Mutant of L1143	1978	Mutation
3.	AAH-1 Desi cotton hybrid	GM sterile DSS / AAH-1	1999	Genetic male (gm) sterility
4.	HHH hybrid	HGMS-1 / HHM-1	2005	gm based *hirsutum* hybrid
5.	Hybrid Kalyan	SH2379 / PIL8	2007	CMS based hybrid
6.	Jawahar Kapas 35	CMSH 6 / RCH2	2007	CMS based hybrid
7.	Orugallu Krishna	WGPH-1 / WGHP-2	2011	Pedigree
8.	Phule Anmol	G. arboretum / G. anomalum	2012	Interspecific hybrid
9.	Phule Dhanwantry	Selection from Nagpur	2012	Mass selection
10.	H 1300	H 1226 / H 1250	2012	Pedigree

3.9 Tea *(Camellia spp.)*

Two species of tea namely, *Camellia sinensis* var. *sinensis*, and *Camellia sinensis* var. *assamica*, are grown widely for beverage in India. Tea is a good example of crops where importance of indigenous material was realised late. Chinese type Tea was introduced in India in the early part of the 19th century. The Assam type tea (*Camellia sinensis* var. *assamica*) was cultivated by natives but its value was realised in 1823 during the Colonial rule. Now it is mostly the Assamese type, which is used commercially not only in India but rest of the world. Tea is an example for yet another plant whose breeding has been done exclusively by private companies. Now tea breeding is done exclusively at Tocklai Experimental Station, Jorhat, which is run by an autonomous organisation, the Tea Research Association.

Tea is a cross-pollinated crop with varying degree of self incompatibility. Clonal selection, varietal crosses, mutation, etc., are used. Seed setting is poor but clonal propagation by means of stem cutting is the main means of varietal multiplication. Some important varieties are listed in Table 14.9. as TV (Tocklai Vegetative) clone and stock (seed produced) types, see Figure 14.3.

Figure 14.3   Mature plant of a hybrid tea variety TV 22 (Courtesy: Dr. I. D. Singh).

Table 14.9   Tea Varieties developed and released in India

Variety	Parentage	Year	Remark
A. Clonal Varieties:			
1. TV 1	19/29/13	1949	Introduction in 1918 and selection
2. TV 2	Stock 14	1959	Introduced in 1914 from China
3. TV 17	TV 1 × St 202	1968	F$_1$ hybrid
4. TV 22	12/41/42	1976	Open pollinated variety
5. TV 24	–	1979	F1 hybrid of Chinese and Assamese type
B. Seed varieties:			
1. Gaurisanker (St. 203)	(4/16, 6/16, 8/7, 8/17, 10/18, 12/15, and 12/18)	1954	Polyclonal cross of 7 clones
2. Nandadevi (St. 378)	(14/5/35) × (14/16/28)	1968	Biclonal F$_1$ hybrid
3. St. 397	(19/29/13) × (19/35/2)	1976	F$_1$ hybrid

4.   USEFUL REFERNCES

Anon, 1960, Cotton in India. Indian Central Cotton Committee, Bombay 402 pp.

-----. 1979. Fifty Years of Agricultural Research and Education. New Delhi : ICAR, 181 PP.

Bhardwaj, D.N. (Ed.) 2012. *Breeding of Field Crops.* Agrobios (India), Jodhpur.

Chopra, V. L. (Ed) 2001. *Breeding Field Crops.* Oxford & IBH Publ. Co. New Delhi, Calcutta.

Dagi, V. (Ed.) 1968. Foundations of Indian Agriculture. ICAR.

Das Gupta, D. 2009. *Sugarcane Development.* Agrobios, India, Jodhpur, Rajsthan, 359 pp.

ICAR 1978. Wheat Research in India 1966-1976, New Delhi: ICAR, 244 pp.

Parthasarathy, N. 1970. Rice Breeding in Asia upto 1960. In : Rice Breeding. Los Banos, Philippines: International Rice Research Institute. Pp. 5-30.

Pushkarnath 1970. *Potato in Sub-tropics.* New Delhi: Orient Longman Ltd. 289 pp.

Ramanujam, S. (Ed.) 1979. Proc. V. Inst. Wheat Genetic Symp., New Delhi, Feb. 23-28, 1978.

Randhawa, M. S. 1963. *Agricultural Research in India: Institutes and Organisation.* New Delhi: ICAR.

Santhanam, V. 1966. *Breeding Procedures of Cotton.* New Delhi: ICAR.

Singh. I. D. 1979. Indian tea germplasm and its contribution to the world's tea industry. Two and A Bud. 26 (9): 23-26.

Singh, S. S.; Sharma, R. K.; Singh, G.; Tyagi, B. S. and Saharan, M. S. 2011. *100 years of wheat Research in India.* Directorate of Wheat Research, Karnal (Haryana), India,281 pp.

Seetharaman, R. and Shobha Rani, N. 1979. High yielding rice varieties in India, their impact and our changing concepts. Ind. J. Agric Sci. 49: 141-150.

Viraktamath, B. C.; Ramesha, M. S.; Hari Prasad, A. S.; Senguttuvel, P.; Raveathi, P.; Kemparaju, K. V.; Shobha Rani, N, and Sailaja, B. 2012. *Two Decades of Hybrid Rice Research in India.* Directorate of Rice Research, Rajendranagar, Hyderabad.

Varietal Release and Seed Production

Manu Smriti (700 - 500 BC) mentions **Subeejam Sukshetre Jayate Sampadyete** (good seed in a good field yields abundant).

1. VARIETY AND STRAIN

A variety is a cultivated strain duly released by an appropriate duly empowered legal organization. After a promising strain or line has been identified by an organization or individual, it is proposed to the empowered legal organization for release. Only when it is officially released it is called variety. Before release it is called a promising strain or line or experimental strain. An experimental strain is either called by a number or its pedigree. After its release it gets a popular name. For example, the rice variety Jaya before its release was called as IET 723. It was thoroughly tested in a number of trials throughout the country and when found superior in yield over other varieties it was released for cultivation under the name of Jaya.

2. RELEASE OF A VARIETY

Formal system of variety release and seed system was formalized under Seed Act 1966, which was repealed by Seed Act 2009. There are two levels of official bodies which are responsible for release of a variety. One is at union or central level and the other at state level. The one at central level is called Central Seed Committee (CSC) based in New Delhi. CSC operates through several sub-committees. **The Central Sub-Committee on Crop Standards, Notification and Release of Varieties of Agricultural Crops** (CSCSNRVA) is the main one concerned with the functions in its name. At state levels there are **State Variety Release Committees** (SVRC) located in most state capitals with the Director of Agriculture as its chairman. Any variety released by SVRC should also be notified by the CSCSNRVA before its seed could be produced and sold.

2.1 State Releases

If a variety is to be released for a particular state, the proposal by the concerned breeder is sent to the chairman of the State Variety Release Committee (SVRC). Usually Director of Agriculture of the State Department of Agriculture is the Chairman of the SVRC. Discussions are held with the members of the committee and the proposing plant breeder to decide whether the proposed experimental strain is suitable for release. If the committee feels satisfied, it is released as a variety, identifying the area of its cultivation. Needless to mention that each state has a specific proforma for submitting the release proposals. But the sum and substance of the proforma is that the morphological characters of plant, seed, yield potential, duration, disease and pest reaction, quality characters, etc, are include in it.

2.2 Central Release

If a variety is suitable for more than one state, it is released by the Central Sub-Committee on Crop Standards, Notification and Release of Varieties of Agricultural Crops (CSCSNRVA). In each crop, there are co-ordinated programmes (see Chapter 14) and during their annual meetings, promising strains are identified and proposed for release.

2.3 Notification

Under the current seed act, each released variety by SVRC or CVRC, must be notified and published in the Government Gazette (http://seednet.gov.in) before its seed can be produced. Therefore, notification proforma is filled and submitted by the proposing organization to the **Central Sub-Committee on**

Crop Standards, Notification and Release of Varieties for Agriculture.
Notification is done only if the committee is satisfied.

2.4 Seed Production

There are basically two types of seed production. The first type is that which
is practised in traditional agriculture. The farmer saves enough out of his best
produce and stores well to use as seed the following year. The second type of
seed production is that which is associated with a progressive modern agriculture.
This type of seed production is based on technical knowledge, equipment and
well defined legal and scientific procedures. Thus it takes into account the
factors related to genetic deterioration of varieties, principles of pure seed
production including certification, processing, storage and distribution. Modern
day plant breeding makes the second type of seed production a necessity.
Still the bulk of seed used by farmers for raising crop is produced by the first
method, but it has to be replaced. The varieties flow through these channels
for production of crops (Figure 15.1).

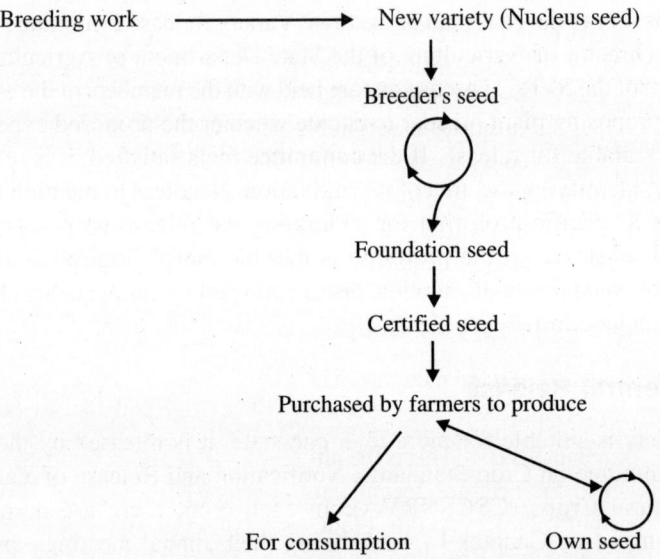

Figure 15.1   Flow of new varieties through the seed system

3. SEED SYSTEM AND CATEGORIES OF SEED

In India Nucleus Seed, Breeder Seed, Foundation Seed, Certified Seed and
Truthful labelled Seed classes (Figure 15.1) are recognized. Nucleus Seed is

the first seed multiplied by the breeder who has developed that variety. It is 100% pure genetically and physically. It is maintained by Maintenance Breeding. Breeder's Seed is utilized to produce Foundation Seed (Figure 15.1) and Foundation Seed is used to produce Certified Seed. Foundation and Certified classes of seeds need certification. Breeder's Seed is only inspected by a team consisting of the breeder, a representative of NSC (not less than Deputy Manager), a representative of certification agency (not less than the rank of Deputy Director of Seed Certification) and the crop co-ordinator (not less than S-2). There is yet another category of seed recognised under the Seed Act. This is called Truthfully Labelled Seed. If any of the above categories of seed has no technical or legal drawbacks and meets the standard of labelling, it may be sold as truthfully labelled seed.

3.1 Seed System

Seed System in India is highly regulated under the Seed Act of 1966 and its subsequent modification by The Seeds Amendment Act 1972, Seed Bill 2004 etc. The seed system in India recognizes Farmers Right and Plant Breeders Right. Farmers are the custodians of the plant genetic resources found in their areas as they are the one who maintained these by growing and caring for it since generatiors. Government of India established a Protection of Plant Variety & Farmers Right Authority based in New Delhi (www.plantauthority.gov.in) in 2001. Plant Breeders Rights are the rights granted by the government to a plant breeder, originator or owner of a variety to exclude others from producing or commercialising a variety for 15–20 years. A person holding PBR can authorise another person or party for it setting the terms.

3.2 Class/Category of Seed

Five categories of seed are recognized in the seed, Nucleus, Breeder, Foundation, Certified and Truthful.

(i) **Nucleus Seed:** This is the ultimate seed stock which breeder maintains out of the original germplasm of the released variety. Nucleus Seed is multiplied by single plant progeny method using the concept of Maintenance Breeding whereby purity of the original parental stock is maintained. Breeder Seed is multiplied out of this material and about 500 single plants are kept annually for continued multiplication as Nucleus Seed.

(ii) **Breeder's Seed:** Breeder's Seed is the seed or vegetative propagated material, which is directly produced by the originating or, in certain cases, the sponsoring breeder or institution. Breeder Seed is multiplied using Nucleus Seed year after year. This step is critical to maintain the original purity and genetic composition of the variety. The cycle of Nucleus Seed and Breeder Seed is maintained using the concept of "Maintenance Breeding" i.e. growing single plant progeny and not practicing simple rouging off type plants but removal of whole single plant-progeny rows, suspecting mutation or outcrossing. This maintenance breeding maintains 100% purity of Nucleus and Breeder Seeds.

(iii) **Foundation Seed:**   This is the progeny of the breeder or foundation seed and is so handled as to maintain the set standards. It is certified by the agency.

(iv) **Certified seed:**   Certified seed is the progeny of foundation or certified seed that is so handled as to maintain satisfactory genetic identity and that has been certified by the agency.

(v) **Truthful Seed:**   Truthful Seed has the same quality standard as the Certified Seed but is not certified by the State Seed Certification Agency. Any organization, including farmers can sell Truthful labelled seed.

Private seed companies are using another category of seed called **Research Seed**. This has the same quality standard as the Certified Seed but this strain is neither released nor certified. Food and Agriculture Organization of the United Nations (FAO) recognizes another category of seed called **Quality Declared Seed** (QDS). This category of seed must meet the seed quality standards of Certified Seed. But QDS is recommended for emergency situation and the producer declares as QDS and is answerable for the quality.

4.   DEGENERATION OF A VARIETY

There are the following reasons for deterioration of a variety, which, if taken care of, are helpful in maintaining the true to the type variety and recommended purity.

4.1   Developmental Variation

When the seed crop is grown in different agroclimates for several consecutive generations, developmental variations may set in. A particular environment may favour some specific type of plants, which may get selected to shift the variety in a particular direction. To reduce the chances of such shifts occurring in varieties, it is advisable to produce their seed in the area of their adaptation.

4.2 Mechanical Mixture

Mixtures may result during sowing, through seed drills, volunteer plants which grow from the remnant seed left in the harvesting machines, at the threshing floor, processing equipment or in storage. Thus mixture is the most important source of varietal deterioration. To avoid such mixtures it is advisable to take precautions during the above operations and follow the principles of seed production.

4.3 Mutation

Spontaneous mutations arise in crop varieties and continuous rouging is the effective way to remove it.

4.4 Natural Crossing

Due to out-crossing (cross pollination) of the variety with another variety, diseased plants and off type plants, the variety deteriorates. The extent of natural crossing depends on the crop, isolation distance, direction and velocity of wind and the presence of pollinators. To avoid varietal deterioration through natural crossing, the recommended isolation distance has to be maintained.

4.5 Minor Genetic Variations

Minor genetic variations (residual variability) may still exist even in the varieties appearing phenotypically uniform and homogeneous at the time of their release. During later cycles of seed production this deteriorates varieties. Following the recommended procedure of seed production helps to overcome this problem.

4.6 Selective Influence of Diseases

Crop varieties gradually become susceptible to obligate parasites and soon go out of the seed programme. Vegetatively propagated crops also deteriorate fast if infected by viral, fungal or bacterial diseases. During seed production it is, therefore, very important to produce disease free stocks and maintain them.

4.7 Techniques of Plant Breeder

Premature release of a variety while still segregating or cytogenetic irregularities may lead to fast deterioration of the variety.

5.   MAINTENANCE OF PURITY IN SEED PRODUCTION

The following steps insure the maintenance of genetic purity in seed production programme:

A.   Genetic Principles

A.1   Control of Seed Source

The use of seed of an appropriate class from an approved source is necessary for raising a seed crop. Breeder's seed is used to produce foundation seed and foundation seed is used to produce certified seed.

A.2   Control of Seed Generation

Genetic purity is further strengthened by the generation system. In this system, seed production is restricted to four generations only. Starting from breeder's seed, the seed can be multiplied only to three more generations through foundation and certified stages.

A.3   Preceding Crop Requirements

This is done to avoid mixture through volunteers. Crops like rice, lentil, etc., have definite requirement of field, which is free of volunteers.

A.4   Isolation

Isolation is kept to avoid contamination through cross pollination, mechanical mixture during sowing to harvesting and spread of seed borne diseases by wind and insects. The isolation distance of various crops has been given in Table 15.1.

A.5   Rouging

Plant in seed fields which differ from the seed variety are referred to as rogue and their removal (rouging) before flowering possibly is a must for seed production.

A.6   Seed Certification

Seed certification is carried out by certification agency to maintain the genetic purity of the crop. During field and seed inspection it is assured whether it meets the minimum standards of quality.

A.7   Growout Tests

Growout test is conducted on the seed lots to be used as seed to ascertain its genetic purity. Usually this is done prior to the main crop season. For example, growout test of seed lots of wheat is done during the summer season in the hills so that their genetic purity may be known prior to sowing during Rabi. Inferior lots, thus, may be rejected.

B. Agronomic Principles

B.1 Suitable Agroclimate

Seed production of a variety should be taken only in the area and season for which it is recommended. The size, shape, appearance of seed lots etc, is best in the suitable agroclimate only.

B.2 Selection of Seed Plot

It should be such that volunteer plants from previous crop season, objectionable weeds, etc., do not pose problems. The location of the plot should be such that isolation distance could be met.

Table 15.1 Isolation distance (meters) of some crops

Crop	Foundation	Certified
1. Rice/barley/oat/wheat/soybean/ground-nut	3	3
2. Pea, mung, urd, lentil, chick pea	20	10
3. Rapes & mustard, Capsicum, Lady's finger	400	200
4. Berseem, lucernes, *Amranthus*	400	100
5. Cauliflower, cabbage, knolkhol, sugar beet, radish, turnip	1000	1000
6. Carrot	1000	800
7. Onion	1000	800
8. Methi, tomato, guar, cowpea, broad bean	50	25
9. Brinjal, pigeon pea	200	100

B.3 Recommended Packages

The recommended package of practices like seed treatment with fungicides, insecticides and Rhizobium culture, time and method of sowing, seed rate, fertiliser, irrigation, weeding, etc, must be followed to insure a healthy crop. Control measures for diseases and insect-pests which affect the seed quality must be applied in time.

B.4 Harvesting

There are two pre-requisites to crop harvesting for seed. First, it must be mature and moisture within the range of recommendation, that is, around 17 per cent; and second, the final inspection by the certification agency must be over and approved. Crop harvesting must not be delayed as it affects the seed quality.

B.5   Threshing and Drying

These operations should be done to insure that there is no varietal mixing and damage to seed. It is advisable to thresh each variety individually. After drying to a recommended moisture level seeds should be put in possibly new bags.

C.   Post Harvest Principles

C.1   Storage of Raw Seeds:

Before and during processing raw seeds are best stored in gunny bags. Bags may be stacked on wooden pallots and only to a height of 3 metres.

C.2   Seed Testing

Seed testing consists of purity, germination and health tests. Only those seed lots are allowed for processing which meet the minimum set standards after seed testing.

C.3   Processing

Processing is totally a mechanical process through human vigil and control to insure proper and mixture free processing. It includes cleaning of seed to remove trash and other undesirable inert matter, grading the cleaned seed into under size, normal size and over size and finally, treating the normal size seed with recommended chemicals like Agrosan GN, Agallol, Thyrom, etc. This treated seed is filled in bags and tagged with recommended tags showing class of seed, year of production, name of variety, germination, purity, etc.

C.4   Storage

The seeds processed in the above way are stored in suitably constructed godowns to protect against dampness, grain pests, rats, etc. a bad storage will spoil the entire care taken from sowing till processing. Therefore storage is an important aspect in retaining germination and vigour of seed lots.

C.5   Seed Legislation

The basic purpose of seed legislation and its subsequent enforcement is to regulate the quality of seed sold to farmers. Until recently (except in Jammu and Kashmir) there was no legislation governing the quality of seed sold to farmers. On 29th December 1966 the Seed Act was passed and it came into force on 2nd October 1969 throughout the country. The main features of this act were the provisions to establish the Central Seed Committee, Central Seed Certification Board, Seed Certification Agencies in various states and Seed

Testing Laboratories in various states. Under the act, notification of varieties, seed law enforcement and penalties for offenders were covered.

The Central Seed Committee advises the States and Central Governments on all matters related to seed, its release, notification, certification standards, etc. The main function of the Central Seed Certification Board is to co-ordinate the work of the State Seed Certification Agencies and insure the uniform application of Seed Certification Standard.

Seed Inspectors appointed by various state governments are responsible for the enforcement of seed law for quality control of seed to be sold to farmers. The inspector can take sample of any seed of a notified variety intended for sale. If on checking the quality is found poor, the person or company is punishable under sections 20 and 21 of the act.

6.   SEED PRODUCTION IN INDIA

Private sector and some government farms were producing seed in their traditional way since long. However, first organized government sector venture started around the time Seed Act was being formulated.

6.1   National Seeds Corporation (NSC)

NSC was formed in 1963 with the idea that it would be the central agency to produce, certify, process and market high quality seed throughout the country. This marked the beginning of systematic seed production based on scientific principles. The headquarter of NSC is New Delhi and it has a number of regional offices in most of the states. With the establishment of National Seeds Project (NSP) the role NSC has changed. Now NSC is responsible for interstate marketing of seed, planning the production of breeder's and foundation seed, production of vegetable and flower seeds of all classes. In states not covered under NSP, NSC continues her original functions that are, production, certification, and marketing.

6.2   State Seed Corporation

As a model Tarai Development Corporation (TDC), now Uttarakhand Seed and Tarai Development Corporation (USTDC), was established in 1969 to the great vision of Dr. Dhyan Pal Singh as an unique organisation in the country to produce, process and market high quality seed. TDC served as a model for developing various state seed corporations under National Seed Project. Success of TDC and quality of Pantnagar seeds set a landmark in the seed industry. The following features made TDC a great success.

1. Involvement of Pantnagar University in technical guidance and supply of breeders' seed, which are essential elements of, seeds programme.
2. Integrated development approach to provide farm machines, land development, irrigation development, electrification, finances to seed producers for raising excellent crop.
3. Growers' participation as shareholders of TDC made them feel responsible for the name, fame and overall growth of the corporation in contrast to the contract system of seed growing followed by NSC.
4. Compact area approach for better technical guidance, supervision, certification, procurement, processing, etc.
5. Strictest quality control through TDC's inspectors, university's laboratory and scientist made Pantnagar seeds set a record.
6. Money back guarantee made it a point to return the price of seed to purchaser if it was found below standard even after sowing.

6.3 National Seed Programme

Following the success of Tarai Development Corporation, Pantnagar, Government of India decided in late 1974 to institute the National Seeds Programme (NSP). Under the aegis of NSP, eight State Seed Corporations were set with the assistance from the World Bank. The principal objective of the NSP is to develop a seed production infrastructure that can keep pace with the rapidly increasing demands for good quality seed of improved varieties. The main features of NSP are:

1. The responsibility of NSP would be shared by various State Seed Corporations for planning and execution of production plans.
2. Facilities of production, processing and quality control would be provided in compact area to produce low cost but high quality seeds.
3. Seed producers would be shareholders of their state corporations.
4. It will strengthen the breeder, foundation and certified seed production and its availability on country basis.

Now most states have their own seed corporations owned by the state government. NSC and the State Seed Corporation backed by the seed legislation of 1966 and modified in 1973, 1981, 2006, 2009 provided policy support for the private sector to grow. Today private seed sector is the major player in supplying, and to some extent in developing, crop varieties in India.

7. SEED CERTIFICATION

The purpose of seed certification is to maintain and make available to the public high quality seeds for growing. Seed testing was started in 1816 in

Switzerland when a law was passed to inspect clover seed before planting. Organised seed testing started in 1869 at Thrandt in Germany under the direction of Friedrich Nobbe. His book Handbuch der Samenkunde (Handbook of Seed Science), published in 1876, was the first book on seed testing in the world. Subsequently seed testing developed in other European countries, USA and elsewhere. In the beginning, the major aim of seed testing was to identify weed seeds which create hazards when present in the crop. But since 1889 seed testing has also included purity and germination tests.

Now there are three international organisations which develop, adopt and publish uniform standards and procedures of seed testing. These are:

International Seed Testing Association (ISTA), established in 1924; Association of Official Seed Analysts (AOSA), established in 1908; and Society of Commercial Seed Technologists (SCST), established in 1922.

In India, seed testing started in the second five year plan (1956-61) with sanction of a grant for establishment of four seed testing laboratories at IARI, New Delhi; Hyderabad in A.P.; Ludhiana in Punjab; and Patna in Bihar. In 1966, a number of laboratories were notified under the Seeds Act, in each state. The IARI laboratory was designated the Central Seed Testing Laboratory.

The Indian Society of Seed Technology was founded in 1971 to organise all those engaged in professions related to seed. ISST publishes two scientific journals, Seed Research and Seed Technology News.

The origin of seed certification is not clear. However, the initial credit goes to Swedish workers who had a system of visiting farmers' fields. To overcome the problems faced in inspection, the International Crop Improvement Association was organised in 1919. Later in 1979 it changed its name to Association of Official Seed Certifying Agencies. In India, the states have their own State Seed Certification Agency to certify seeds.

In most states certification is done by the State Seed Certification Agencies. The National Seed Corporation continues to certify seeds where state agencies do not exist. Seed certification is a legally sanctioned system for quality control of seed production which consists of the following:

1. **Seed Source:** An administrative check on the origin of seed used for seed multiplication.

2. **Field Inspection:** Evaluation of growing crop for isolation, mixtures, objectionable weeds and diseases.

3. **Procedure of Production:** To ensure that the grower is following the recommended package of practices to get a healthy crop, mixture free and healthy seeds.

4. **Sample Inspection:**   To check if the seed lot meets the minimum seed standards.

5. **Growout Test:**   To check the purity and quality of various seed lots. Steps 1-3 are called field inspection and 4-5 seed inspection.

7.1 Procedure of Certification

(i) **Field Inspection:**   The seed certification agency, on receipt of application for registration from the grower, shall examine it for source seed used for sowing, etc., and then shall register the grower. For field inspection a notice is given to grower about the time of inspections. Usually 2-4 inspections are done from preflowering to maturity phase depending on the crop species, and isolation distance, crop condition, off types, objectionable weeds, etc., are observed. A copy of each inspection report is given to the grower with a specific remark if the crop meets the field standards. the field standards for some crops are given in Table 15.2. Isolation distances have already been given in Table 15.1

In the final inspection, which is done at maturity, if the crop is found to meet the standards, it is accepted as seed crop for foundation or certified seed class.

The objectionable weeds are those which are difficult to eradicate, poisonous, or whose seeds are inseparable from the crop. For example, wild rice or red rice for rice, Kasni for berseem, Cuscuta for Lucerne, *Argemone mexicana* for rapes-mustard are objectionable weeds. Diseased ear is counted for designated diseases' which are seed borne and detrimental to crop. For example, loose-smut for wheat, leaf roll and scab for potato are designated diseases.

(ii) **Seed Inspection:**   After a certified crop is harvested and threshed, grain samples are drawn to test if these meet the minimum quality seed standards (Table 15.3).

The detail method of seed inspection which is based on the test conducted by seed testing organisations is discussed in the next section, seed testing.

(iii) **Seed Testing:**   Seed testing is an integral part of seed production to control its quality. It has following objectives:

1. To determine the suitability of seed for sowing.
2. To guide the seed industry about procedures of drying and processing so as to obtain the best quality seeds.

Table 15.2. Minimum field standards for certification of foundation (F) and certified (C)

Crop	Off type %		Inseparable crop plants %		Objectionable weeds %		Diseased ear %	
	F	C	F	C	F	C	F	C
1. Barley, rice, oat, wheat	0.05	0.05	0.01	0.05	0.01	0.02	0.1	0.5
2. Bajra	0.01	0.05	–	–	–	–	0.5	0.1
3. Rapes	0.1	0.5	–	–	0.05	0.1	0.1	0.2
4. Jute	0.5	1	–	–	–	–		
5. Gram, chickpea, soybean, urd	0.1	0.2	–	–	–	–	–	–
6. Cole crops	0.1	0.5	–	–	–	–	0.1	0.5
7. Tomato	0.1	0.5	–	–	–	–	0.1	0.5
8. Berseem, lucernes	0.2	1	–	–	None	0.05	–	–

Table 15.3 Minimum seed standards for Foundation (F) and Certified (C) seeds (Values in per cent except some of columns 3, 4, and 5)

Crop	1*	2	3		4		5		6	7	8
	F&C	F&C	F	C	F	C	F	C	F&C	F&C	F&C
Rice	98	2	10/kg	0.1	10/kg	0.1	2/kg	5/kg	80	13	8
Wheat	98	2	10/kg	0.1	10/kg	0.1	2/kg	5/kg	82	12	8
Sorghum	98	2	5/kg	0.1	5/kg	0.1	--	--	80	12	8
Bajra	98	2	10/kg	0.1	10/kg	0.1	--	--	75	12	8
Cauliflower	98	2	0.05	0.1	.05	0.2	--	--	65	7	5
Rapes	97	3	0.1	0.5	0.1	0.5	0.05	0.1	85	8	5
Tomato	98	2	0.05	0.1	None	None	--	--	70	8	5
Berseem	98	2	0.1	.5	0.1	0.5	5/kg	20/kg	80	10	7

* 1 = Pure seed (minimum), 2 = Inert mater (maximum), 3 = Other crop seed (maximum), 4 = Weed seeds (maximum), 5 = Objectionable weed seeds (maximum), 6 = Germination (minimum), 7 = Moisture in ordinary container, 8 = Moisture in airtight container.

3. To guide farmers for discrimination among seed lots in the market based on quality.
4. To check if the seed lot meets established quality standards and label specifications.
5. To identify problems of seed quality in seed production.

The following types of tests are performed on the seed samples under seed testing:

1. Seed moisture
2. Physical purity
3. Genetic purity
4. Germination
5. Seed vigour
6. Seed health

8. PROCEDURE OF SAMPLING AND SEED TESTING

Samples are drawn from the stored seed lots by taking small portions at random (primary sample) from different portions in the lot. Primary samples taken from the same lot are combined to make composite samples from which small samples (submitted samples) are drawn to be supplied to the laboratory. Working samples are drawn out of the submitted samples for various tests in the laboratory. Flow chart of samples for various tests is shown in Figure 15.2.

The number of primary samples depends on the size of the seed lot and crop. Various types of seed triers; thief, sleeve, bin, noble (Figure 15.3) are used to take out samples from bags, bins, heaps, etc. the size of the submitted sample finally depends on the crop. For example, 15 gms for tomato, 25 gms for tobacco, 150 gms for muskmelon, cucumber, okra, brinjal, chillies, linseed, jute, bajra, 500 gms for rice, sugar beet, and 1000 gms for wheat, maize, oats, barley, pigeon pea, chickpea, beans, urd, mung, groundnut, castor, pea, etc.

From the submitted samples, working samples are drawn in the laboratory using a mechanical divider (Boerner or Gamet type, Figure 15.4), by the random cup method or the spoon method. The basic principle in these methods is that a representative sample is drawn for analyses.

8.1 Purity Analysis

The working sample is weighed and separated into the following three groups by weight and expressed in per cent:

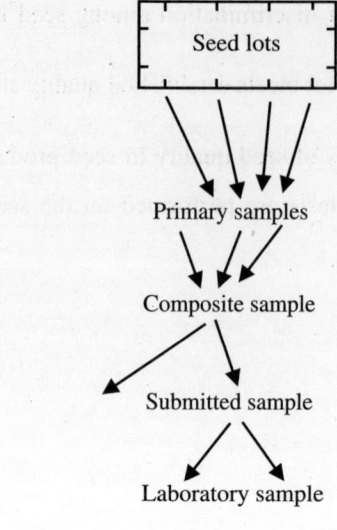

Figure 15.2 Various types of seed samples and tests conducted.

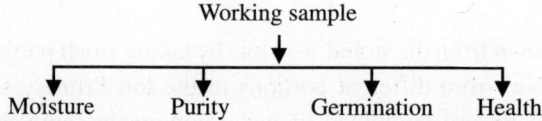

Figure 15.3 Seed samplers: A = for heaps and godowns; B = for Bags.

(a) **Pure seed:** This includes intact seeds of that species, broken seeds if more than half of size, (seeds of Leguminoseae, and Cruciferae without seed coat are regarded as inert matter).

(b) **Other seeds:** This group shall include seed and seedlike structures of any plant species other than of pure seed.

(c) **Inert matter:** This shall include other matter (soil clods, sand, chaff, stems, leaves, etc., ergot sclerotia, bunt balls, insect larvae) and seed

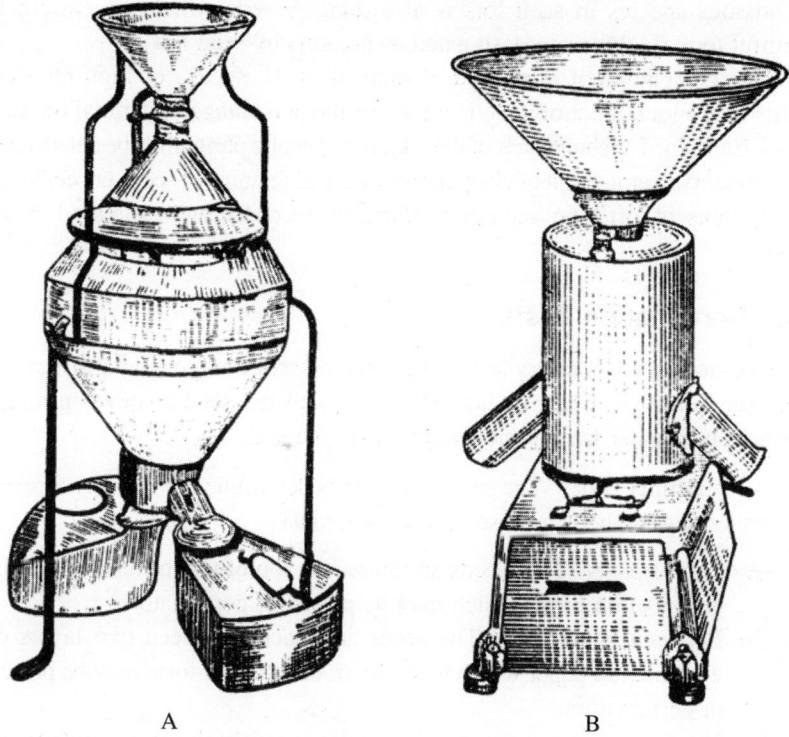

A   B

Figure 15.4 Mechanical Dividers: A = Boerner type; B = Gamet type

and seedlike structures (less than half portions of seeds, coatless seeds of the above two families).

Figure 15.5 Purity Work Board.

Analyses are done by placing the weighed sample on purity work board (Figure 15.5) and separating it into the above three groups. These three components are expressed in per cent. Along with purity analysis, presence of noxious species in seed lots is also done. Weeds which are extremely harmful to agriculture are designated as noxious by seed law.

The percentage of pure seed, if more than 98, speaks of high physical purity of the lot but not of genetic purity, as the percentage of varietal mixture is not determined. Genuineness of the variety or genetic purity may be determined by laboratory examination of characteristic varietal features of seeds or seedlings. The number of 'off-type' seed or seedling are recorded and expressed in per cent.

8.2 Germination Tests

Germination, in a laboratory test, is the emergence of those essential structures from seed embryo, which indicate the ability of the seed to develop into a normal plant under favourable conditions in the soil.

A. **Viability test:**   The germination tests determine the viability of seed. The following methods are used for viability test.

(a) Top of paper (TP). Seeds are placed on top of one or more layers of germination paper which may be placed in germinators.
(b) Between paper (BP). The seeds are placed between two layers of germination paper which is rolled, or the straight form may be placed in germinators.
(c) Using sand or soil (S). Seeds may be germinated using sand or soil. Seed may either be placed on top of sand or soil or sown inside it.

The requirement of specific crop for germination is different. Some seeds require light, others require a pre-treatment to break dormancy. This is done as per requirement to test the germination. The requirement of the temperature is also variable. With some examples these points are given in Table 15.4.

Table 15.4   Method of conducting germination tests of some crops

Crop	Method	Temperature	First count (days)	Final count	Remarks
Rice	BP, TP	20-30°C	5	14	Pre -soak 24-48 hrs.
Wheat	S, BP	20°C	4	8	Diffuse light
Rapes	TP, BP	15–25°C	3	7	Light, Prechill
Jute	TP, S	30°C	3	5	Full light
Melons	BP, S	20–30°C	4	8	Low moisture
Okra	BP, TP, S	20–30°C	4	21	None

After the final count, seedlings are classified into normal seedlings (having well branched root with root hairs, and well developed plumule) and abnormal seedlings (deformed, deficient and decayed seedlings). The seeds which did not germinate are classed into three groups.

   (i) **Hard Seeds:** Seeds of Leguminoseae and Malvaceae which remain hard due to not absorbing water;

 (ii) **Fresh ungerminated seeds:** Seeds other than hard seeds which remain firm and apparently viable, and

(iii) **Dead seeds:** Seeds which are neither hard nor fresh. The per cent of normal seedlings is taken as true as germination percentage.

Germination tests require long time and hinder the work of processing and marketing. This necessitated the development of rapid methods of estimating germination of seed lots. There are two acceptable methods, which are called Rapid Germination Method:

 (a) **Tetrazolium test:** The embryo is exposed by cutting the seed and stained in 1 to 2 per cent solution of 2, 3, 5 triphenyl tetrazolium chloride. The live cells take the dark red stain and are evaluate under a magnifying glass or microscope. The embryos of dead seeds do not take stain and remain colourless.

 (b) **Embryo excision method:** The embryo is excised out of soaked seed sample and incubated. The viable ones start differentiation.

8.3   Seed Vigour Tests

The objective of germination test in a seed testing laboratory is to evaluate the viability or stand-producing potential of seed lots. In many instances, however, it has been noticed that seed lots of similar germination percentage gave varying plant population in the field. This is primarily due to the fact that the viability test does not specify if a germinating seed in the laboratory would produce a plant under field conditions. The stand producing potential of a seed depends on its vigour. There are two major methods in use to test it:

 (a) **Indirect test:** Indirect test includes the measurement of seedling length, dry matter produced in five weeks or speed of germination of various lots. It is assumed that seeds which grow faster initially, or produce more dry matter on germination, or germinate more quickly are more vigorous.

 (b) **Direct method:** Direct methods include direct testing by brick-gravel test and paper piercing test. In brick-gravel test, the seed covered with 30 mm layer of brick-gravel of 3 mm size. Only vigorous seedlings emerge out and weaker ones remain underneath. Similarly a paper sheet of 0.4 mm and 90 gm/m^2 weight is placed above the germinating seed. Only vigorous seedlings come out piercing the paper sheet.

8.4   Seed Moisture Testing

Moisture of seed is the most important factor influencing the retention of viability and appearance of seed. Therefore, in seed lots a maximum limit of moisture is permitted for certification and storage (see columns 7 and 8 of Table 15.3).

There are a number of methods to determine the moisture in the sample, such as drying with P_2O_5, freeze drying, infra-red moisture meter, Karl-Fischer titration method, oven drying, Toluene distillation method, Nuclear magnetic Resonance and electric moisture meters. Most of them are either time consuming, require sophisticated equipments and chemicals and are thus impracticable. Universal moisture testes and OSAW moisture tester used commonly in India. These are reasonably precise and pretty fast.

8.5   Seed health Test

The health of the seed primarily refers to the presence or absence of disease causing organisms, such as fungi, bacteria, viruses, nematodes, insect pests, etc. Information on seed health is important for the following three reasons:

1. Seed borne inoculum may give rise to progressive disease development in the field.
2. Foreign introductions may introduce new diseases.
3. It may elucidate seedling evaluation and causes of poor germination or field stand.

Method of testing: The presence or absence of the pathogen of a designated disease may be determined by any of the following methods depending on the disease.

(a) Direct examination of ergots, bunts, smuts, discoloured grains and insect damage is made on 400 seeds of each lot. Alternatively, the imbibed seed or seed washing are observed for pathogen under the microscope.
(b) Seeds may be placed on blotter, sand or agar and incubated to make the pathogen more visible. Observations may then be recorded.
(c) Specialised techniques like growing plants, serological reactions as in the case of potato for virus, or observation on separated embryo are applied.

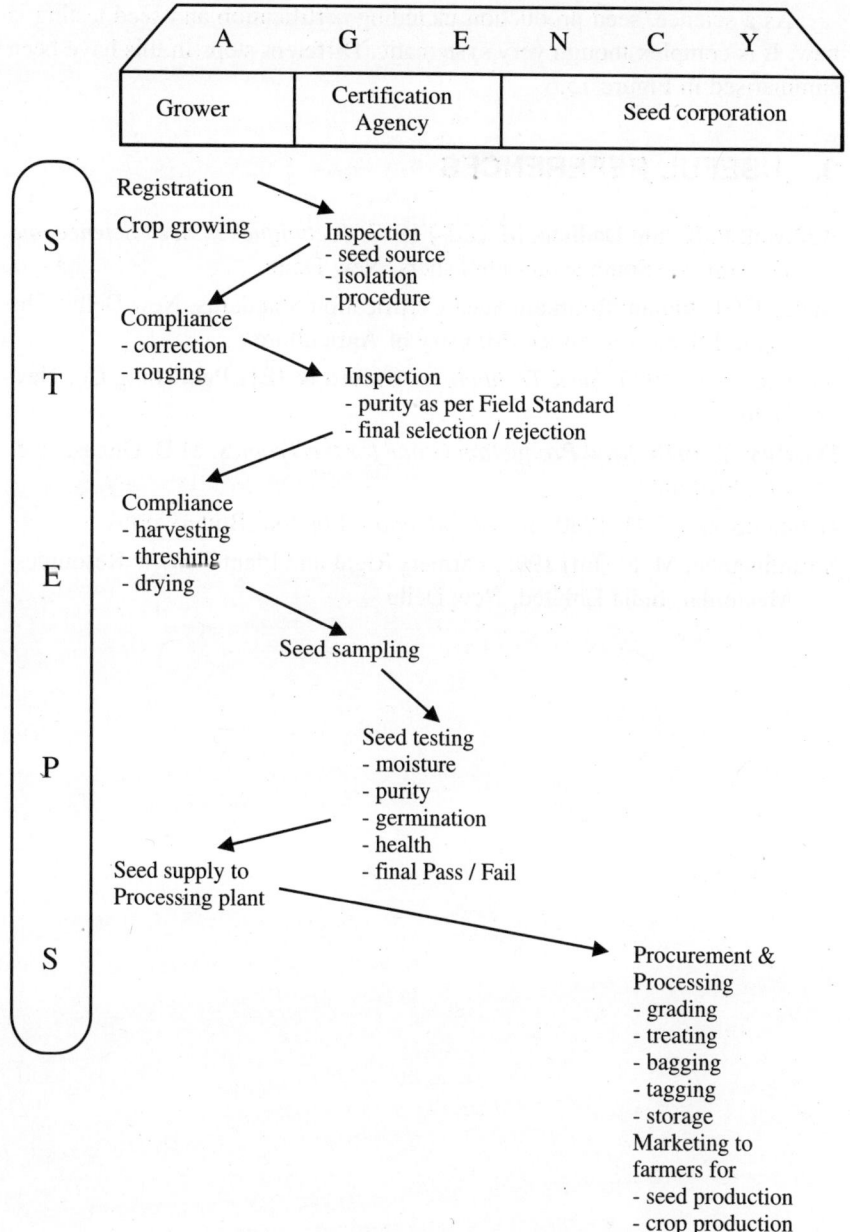

Figure 15.6 Flow chart of seed production and agencies involved in it. The arrow stops if the previous step was not o.k.

As a science, seed production including certification and seed testing is new. It is complex though very systematic. Different steps in this have been summarised in Figure 15.6.

9. USEFUL REFERENCES

Agrawal, P. K. and Dadlani, M. (Eds) 1987. *Techniques in Seed Science and Technology*. South Asian Publishers, New Delhi.

Anon. 1971. Indian Minimum Seed Certification Standards. New Delhi: The Central Seed Committee, Ministry of Agriculture.

Agrawal, R.L. 1995. *Seed Technology*. Oxford & IBH Publishing Co. New Delhi.

Doerfler. T. 1976. *Seed Production Guide for the Tropics*. M.D. Gunasena & Co., Sri Lanka:

Hebblethwaite, P.D. 1980. *Seed Production*. London: Butterworths.

Swaminathan, M. S. (Ed) 1995. Farmers Right and Plant Genetic Resources. Macmillan India Limited, New Delhi.

Glossary

Acclimatization: The process that leads to the adaptation of newly introduced variety to a new environment is called adaptation. Some degree of genetic variability must be present in the introduced variety for the acclimatization to occur.

Adaptation: The process by which individuals (or parts of individuals), populations, or species change in form or function in such a way to better survive under given environmental conditions. Also the result of this process is also called adaptation.

Addition Line: An addition line has one pair of chromosome from another variety or species in addition to the normal somatic chromosome complement (2n) of the species.

Allele: Allele is one of the two or more alternate forms of a gene occupying the same locus on a particular chromosome. Members of a set of alleles are mutually exclusive, arise by gene mutation and their activity is concerned with the same biochemical and developmental process.

Allopolyploid: A polyploid containing genetically different sets of chromosomes, for example, sets from two or more different species.

Amphidiploid: A polyploid whose chromosome complement is made up of the entire somatic complements of two diploid species.

Aneuploid: An organism whose somatic chromosome number is not an even multiple of the haploid number.

Antibiosis: It refers to an adverse effect of feeding on an insect, on a resistant host plant, on the development and or reproduction of an insect-pest.

Apomixis: Reproduction in which sexual organs or related structure take part but fertilisation does not occur, so that the resulting seed is vegetatively produced. If a hybrid is apomictic, it breeds true and no segregation occurs and is called "permanent hybrid:.

Asexual Reproduction: Reproduction which does not involve the union of male and female gametes. The plant develops either from vegetative parts or from reproductive parts without fertilization.

Autogamy: Self-fertilisation.

Autopolyploid: A polyploid arising through multiplication of the complete haploid set of a species.

Backcross: A cross of a hybrid to either of its parents. In genetics, a cross of a heterozygote to a homozygous recessive. (See test cross).

Backcross Breeding: A system of breeding whereby recurrent backcrosses are made to one of the parents of a hybrid, accompanied by selection for a specific character or characters.

Balanced Lethal Sysetm: A genetic system in which two recessive lethal genes are linked tightly in repulsion phase e.g. l_1L_2 / L_1l_2 and only the heterozygotes can survive.

Basic Number: The number of chromosomes in diploid ancestors of polyploids, represented by x.

Biometry: The branch of science which deals with statistical procedures in biology.

Biotype: A group of individuals with the same genotype. Generally used in reference to insect pests as a race.

Bivalent: A pair of homologous chromosomes united in the first meiotic division.

Boom and Bust Cycle: When a new variety with a vertical resistance gene is developed to the prevalent pathotype of a pathogen it gets cultivated in large area leading to "boom period". Once it becomes susceptible, area under it reduces leading to "bust period". Thus varieties with vertical resistance gene show Boom and Bust cycle.

Breeder Seed: Progeny of Nucleus Seed produced by the agency which has developed that variety or authorised by the developer is called Breeder Seed. Breeder Seed is used for producing Foundation Seed.

Bulk Breeding: Growing of genetically diverse populations of self-pollinated crops in a bulk plot with or without mass selection, followed by single-plant selection.

Certified Seed: Progeny of Foundation Seed, which is certified by a certification agency and meets the set field and laboratory standards.

Chiasma: An exchange point of chromosome between paired chromatids in the first division of meiosis.

Chromatid: One of two threadlike structures formed by the longitudinal division of chromosome during meiotic prophase and known as a daughter chromosome during anaphase.

Chromosomes: Structural units of the nucleus which carry the genes in linear order.

Chromosome Elimination: In cases of distant or wide hybridisation, some chromosomes get gradually eliminated from the zygote. Embryo develops normally but plants obtained do not have full chromosome compliments of the two parents and are not true interspecific hybrid.

Clone: A group of organisms descended by mitosis from a common ancestor. Clonal population is supposed to have heterozygous individual but homogenous population.

Combining Ability: Average performance of a strain in a series of crosses is called general combining ability. Specific combining is the performance of a hybrid combination in specific cross predicted as deviation from general combining ability.

Composite: Composites are varieties produced by mixing the seeds of several phenotypically outstanding lines and encouraging open pollination to produce crosses in all combinations. However, the combining ability of lines is not tested.

Coupling: Linked recessive alleles occur in one homologous chromosome and their dominant alternatives occur in the other chromosome. Opposed to repulsion in which one dominant and one recessive occur in each member of the pair of homologous chromosomes.

Covariance: The mean of the product of the deviation of two varieties from their individual means. A statistical measure of the interrelation between variable.

Crossing Over: The exchange of corresponding segments between chromatids of homologous chromosomes during meiotic prophase. Its genetic consequence is the recombination of genes.

Cybrid: These are hybrid cells containing nucleus of one species but cytoplasm from another species produced in normal protoplast fusion experiments.

Cytoplasmic Inheritance: Transmission of characters through cytoplasm. It is due to the DNA present in the cytoplasmic organelles.

Cytoplasmic-genetic Male Sterility: Cytoplasmic male sterility for which a restorer gene is known or available.

Cytoplasmic Male Sterility: Male sterility showing cytoplasmic inheritance.

Detassel: Remove the tassel (male inflorescence) as in maize.

Diallel Selective Matiing Scheme: Crossing of selected individual plants in a segregating population, say F2, F3 etc., and allowing these to segregate further for developing purelines in self pollinated crops. Jensen proposed his scheme in 1970.

Dihybrid: Heterozygous with respect to two genes.

Diploid: An organism with two chromosomes of each kind of a haploid set.

Distant Hybridisation (Wide Hybridisation): Hybridization between individuals belonging to same genus or different genera.

Donor Parent: The parent from which one or a few genes are transferred to the recurrent parent in backcross breeding. Donor parent is used for transferring disease resistance, quality or any specific character.

Double Cross: A cross between two F1 hybrids.

Drift: Changes in gene and genotypic frequencies in small populations due to random processes.

Duplication: Occurrence of a segment of a chromosome twice in the haploid set.

Emasculation: Removal of the anthers (male part) from a bisexual flower.

Electroporation: Introduction of DNA into cells or protoplasts by exposing them to high voltage current for a few milliseconds. This opens transient micropores in plasmalemma and promotes exchange of DNA.

Embryo Rescue: Use of embryo culture to give life to an abortive embryo in distant cross, which otherwise would die due failure of endosperm development.

Epistasis: Dominance of one gene over a non-allelic gene. The gene suppressed is said to be hypostatic. In population and quantitative genetics, epistasis refers to all types of non-allelic interactions or intra-allelic interactions.

Environment: The sum total of the external conditions which affect growth and development of an organism. Environment may consists of soil, water, temperature, humidity, light etc.

Epiphytotic: Artificial creation of heavy disease pressure / pest pressure (epidemy). This is done to screen germplasm or breeding lines for identify resistance donors.

Epidemic: An un-arrested spread of a plant disease over a wide area.

Expressivity: The degree of manifestation (phenotypic expression) of a genetic factor or a gene.

F1: The first generation of a cross.

F2: The second filial generation obtained by self-fertilisation or crossing inter se of F1 individuals.

F3: The third filial generation obtained by self-fertilising F2 individuals.

Factor: Same as gene.

Gene: It is unit of inheritance and in genetic terminology defined as particular sequence of nucleotides along a molecule of DNA. Genes are located at fixed loci in chromosomes and can exist in a series of alternative forms called alleles. These could exist as dominant, recessive, co-dominant or multiple alleles. Structurally, gene is a segment of DNA, which codes for one polypeptide, ribosomal or transfer RNA.

Gene Bank: A facility where large number germplasm accessions of organisms are stored in short, medium or long terms storages.

Gene For Gene Relationship: Flor in 1956, based on his work on linseed rust, postulated that for every resistance gene present in the host, pathogen has a gene for virulence.

Gene Frequency: The proportion in which alternative alleles of a gene occur in a population

Gene Pool: Sum total of all genes present a population of an individual organism.

Germplasm: The sum total of the hereditary materials present in a species.

Genome: A set of chromosomes corresponding to the haploid set of the species.

Gene Pool: The total genetic information encoded in the sum total of the genes in a breeding population existing at a given time is known as gene pool. The gene pool and genotype frequencies may remain unchanged in absence of migration, mutation and selection.

Genetic Drift: A change either directed or undirected in gene frequency in a population is called genetic drift.

Geetic Engineering: The genetic manipulations by which an individual having a new combination of inherited properties is established is called genetic engineering. It may be done either by cellular manipulation of cell hybridisation or molecular manipulation of inserting alien genes in host DNA.

Genetic Equilibrium: The situation in which both the gene and genotype frequencies in a large randomly mating population remain constant through successive generations.

Genetic Erosion: Loss of genetic variability in the local germplasm (variety) due to few improved varieties dominating in the cultivation.

Genetic Map: Genetic map is the representation of the genetic distance separating non-allelic gene loci in a linkage structure.

Genome: Genome consists of complete chromosome set of an organism consisting of a species specific linkage groups, hence the sum total of its genes.

Genotype: The entire genetic constitution of an organism.

Haploid: A cell or organism with the gametic chromosome number (n).

Heritability: The proportion of observed variability which is due to heredity, the remainder being due to environmental causes. More strictly, the proportion of observed variability due to the additive effects of genes.

Heterosis: Hybrid vigour such that an F1 hybrid falls outside the range of the parents with respect to some character or characters.

Heterocarysis: The presence of two or more genetically different nuclei within single cells of a mycelium.

Heterothally: Haploid incompatibility in fungi (opposite of homothally).

Heterozygous: Having unlike alleles at one or more corresponding loci (opposite of homozygous).

Homozygous: Having like alleles at corresponding loci on homologous chromosomes. An organism can be homozygous at one, several, or all loci.

Horizontal Resistnace: Resistance usually governed by polygenes, and is race non-specific.

Hybrid: The product of a cross between genetically dissimilar parents. The parents may differ in single or multiple genes.

Ideotype: An ideotype has all the characteristics considered ideal for a plant under a given environment to yield the highest. For example ideotype for rice and wheat are semi-dwarf stature, non-lodging string stem, erect leaves etc.

Inbred Line: A line produced by continued inbreeding. In cross pollinated crops inbreds are developed using self pollination followed by selection once reasonable homozygosity has been achieved.

Inbreeding: The mating of individuals more closely related than individuals mating at random.

Inbreeding Depression: Inbreeding depression is the reduction or loss of vigour and fertility due inbreeding.

Irradiation: Exposure of plants or plant parts to X-rays or other radiations to increase mutation rates.

Isogenic Lines: Two or more lines differing from each other genetically at one locus only. Distinguished from clones, homozygous lines, identical twins, etc., which are identical at all loci.

Isolation: The separation of one group from another so that mating between or among groups is prevented.

Karyotype: The particular chromosome complement of an individual as defined by number and morphology of the chromosome usually in mitotic metaphase.

Linkage: Association of characters in inheritance due to location of genes in proximity on the same chromosome.

Linkage Group: Linkage group is a group of gene loci, which can be placed in a liner order representing the different degrees of linkage among the genes concerned. Number of linkage group is limited to the number of chromosomes per genome i.e. haploid number of chromosomes in a diploid organism.

Male Sterility: Absence or nonfunction of pollen in plants.

Mass Selection: A form of selection in which individual plants are selected and the next generation propagated from the aggregate of their seeds.

Mating System: Any of a number of schemes by which individuals are assorted in pairs leading to sexual reproduction. Random, assortment of pairs by chance. Genetic assortative mating, mating together of individuals more closely related than individuals mating at random. Genetic disassortative mating, mating together of individuals less closely related than individuals mating at random. Phenotypic assortative mating, mating individuals more alike in appearance than the average. Phenotypic disassortative mating, mating of individuals less alike in appearance than individuals mating at random.

Metaxenia: Influence of pollen on maternal tissue of the fruit. (See xenia).

Monoecious: Staminate and pistillate flowers borne separately on the same plant.

Mutlilines Variety: Mixture of several purelines released as a variety. This is usually done to control a disease or adverse environment..

Multiple Allele: A member of a series of more than two alternative forms of a gene.

Mutagen: A chemical or physical agent used for inducing mutation.

Mutation: A sudden heritable variation in a gene or in chromosome structure.

Non-recurrent Parent: In back crossing breeding, the parent which is used only once in crossing to transfer one or a few desirable genes.

Notification of a Variety: After a crop variety is realised **The Central Sub-Committee on Crop Standards, Notification and Release of Varieties of Agricultural Crops** considers it. If satisfied, the name and details of the variety are published in the Government Gazette, and then only the variety enters in the seed production system.

Nullisome: An otherwise 2n plant that lacks both members of one specific pair of chromosomes, hence, with 2n--2 chromosomes.

Oligogenes: Also called major genes, these have major individual effect on the phenotype. This is in contrast to polygenes or minor genes, which have small individual effect.

Parthenogenesis: Development of an organism from a sex cell but without fertilisation.

Pedigree: A record of the ancestry of an individuals, family, or strain.

Pedigree Breeding: A system of breeding in which individuals plants are selected in the segregating generations from a cross on the basis of their desirability judged individually and on the basis of a pedigree record.

Penetrance: The degree with which a gene produces a recognisable effect in individuals which carry it.

Phenotype: Appearance of an individual as contrasted with its genetic make-up or genotype. Also used to designate a group of individuals with similar appearance but not necessarily identical genotypes.

Physiological Races: Pathogens of the same species with similar or identical morphology but differing in pathogenic capabilities.

Polycross: Open pollination of a group of genotypes (generally selected) in isolation from other compatible genotypes in such a way as to promote random mating inter se.

Polygenes: Genes whose effects are too slight to be identified individually but which, through similar and supplementary effects can have important effects on total variability.

Progeny Test: A test of the value of a genotype based on the performance of its offspring produced in some definite system of mating.

Propagule: Plant pat used for propagation of the plant, may be seed or any other vegetative part

Pure Line: Strain homozygous at all loci, ordinarily obtained by successive self-fertilisations in plant breeding.

Qualitative Character: A character in which variation is discontinuous.

Quantitative Character: A character in which variation is continuous so that classification into discrete categories is not possible.

Random Drift: It is a random change in gene frequency due to sampling error.

Random Mating: In a random mating, each female gamete is equally likely to unite with any male gamete and the rate of reproduction of each genotype is equal.

Reciprocal Crosses: Crosses in which the male and female parents are reversed.

Recurrent Parent: The parent to which successive backcrosses are made in backcross breeding starting with F1 itself.

Recurrent Selection: A method of breeding designed to concentrate favourable genes scattered among a number of individuals by selecting in each generation among the progeny produced by matings inter se of the selected individuals (or their selfed progeny) of the previous generation.

Rogue: A variation from the standard type of a variety or strain. Rouging is removal of undesirable individuals to purify the stock.

Self-incompatibility: Genetically controlled physiological hindrance to self-fruitfulness.

Sibs: Progeny of the same parents derived from different gametes. Half sibs, progeny with one parent in common.

Single Seed Descent Method: It is a modification of bulk method in which a single seed from each F2 plant is bulked to raise F3 generation. This may continue in later generation till selection can effectively be done. This method has disadvantage of multiplying undesired segregants at the cost of desirable ones.

Strain: A group of similar individuals within a variety.

Substitution Line: A line in which a pair of chromosomes has been replaced by a pair from another variety of the same species.

Systhetic Variety: A variety produced by crossing inter se a number of genotypes selected for good combining ability in all possible hybrid combinations, with subsequent maintenance of the variety by open pollination. Thus Synthetic Variety is slightly different from Composite Variety.

Test Cross: A cross of a double or multiple heterozygote to the corresponding multiple recessive to test for homozygosity or linkage.

Transgene: An alien gene introduced into an organism by means of genetic engineering. This gene may be from the same organism or different organism or a synthetic sequence.

Transgenic: An individual having gene transferred from unrelated species or organism. This is done using biotechnological tools or genetic engineering. The individual with trangene or transgenic is called genetically modified organism (GMO).

Transgressive Segregation: Appearance in segregating generations of individuals falling outside the parental range in respect to some character.

Variance: Mean squared deviation of a population of variates from their mean the square of the standard deviation. The corresponding statistic is the mean square.

Variation: The occurrence of differences among individuals due to differences in their genetic composition and /or the environment in which they were raised.

Vertical Resistnace: Race specific resistance, usually controlled by major genes, against a race or pathotype.

Vertifiloia Effect: It is the loss of resistance due to appearance of virulent pathotypes against major resistance genes. The name is derived from the German potato variety Vertifiloia whose R_3 and R_4 resistance genes succumbed completely to new virulent pathotypes P_3 and P_4.

Wide Cross: Cross between two species of the same genus or of different genius. This is also called distant hybridisation.

Xenia: Effect of genotype of pollen grain on the phenotype of embryo and endosperm.

Subject Index